Using Assessment to Improve Middle-Grades Mathematics Teaching & Learning

Suggested Activities Using QUASAR Tasks, Scoring Criteria, and Students' Work

Carol S. Parke
Duquesne University
Pittsburgh, Pennsylvania

Suzanne Lane
University of Pittsburgh
Pittsburgh, Pennsylvania

Edward A. Silver
University of Michigan
Ann Arbor, Michigan

Maria E. Magone
University of Pittsburgh
Pittsburgh, Pennsylvania

CONTENTS

Contents of the CD

ACKNOWLEDGMENTS

We would like to acknowledge the efforts of the following people who have contributed over the years to the QUASAR Cognitive Assessment Instrument (QCAI), from the development of tasks and rubrics to the validation of the assessment instrument as a whole: Robert D. Ankenmann, Jinfa Cai, Adam Deutsch, Connie Finseth, Mary Jakabcsin, Mei Liu, David Meel, Barbara Moskal, Clement A. Stone, Ning Wang, and Yuehua Zhu.

We are also grateful to three other QUASAR team members, Margaret S. Smith, Mary Kay Stein, and Marjorie Henningsen, who were instrumental in providing support to the teachers and resource partners as they used the QCAI release tasks in their classrooms and in professional development sessions. The thorough documentation of these interactions and others provided us with the rich information that formed the basis of some of the activities included in this book.

Finally, a special thanks goes to all the QUASAR teachers and resource partners at the six school sites. Without their use of the tasks, student responses, and scoring materials, this book would not have been possible.

INTRODUCTION

"THE MORE I incorporate these kinds of tasks in my classroom and we talk about the responses and evaluate them, the more adept my students become in problem solving and reflecting on their work."

"I know more about my students. I used to settle for an answer, and now, they have to explain and I can understand what is going on in their heads…. I used to feel, 'This is my room and the floor is my water, and I'm the only one in here who can walk on water.' Now I let them walk on water, too, and share my floor."

"To me as a teacher, there's a lot of value in looking at students' explanations. We get a more complete picture of what our students have learned and how they are thinking about a particular math concept. I feel that I'm better able to make instructional decisions based on this information."

Such comments are typical when teachers use tasks that require students to think deeply about problem situations and communicate their mathematical understanding rather than merely apply procedures mechanically. In fact, these sentiments echo *Principles and Standards for School Mathematics (Principles and Standards)* (NCTM 2000), which notes that assessment should enhance student learning and can be a valuable tool for making instructional decisions. Further, *Principles and Standards* states, "To maximize the instructional value of assessment, teachers need to move beyond a superficial 'right or wrong' analysis of tasks to focus on how students are thinking about the tasks" (p. 24). Using scoring criteria in rubrics not only helps teachers determine students' levels of proficiency but also allows them to gain insight into students' mathematical thinking. Classroom discussions of scoring criteria also help students become aware of various approaches to solving tasks and to begin to understand the characteristics of a complete and correct response to a complex problem.

This book and companion CD include a set of mathematics tasks and related scoring materials that can be used by classroom teachers and mathematics educators who work with teachers for a variety of instruction and assessment purposes. Specifically, the book and CD contain a total of sixteen mathematics tasks and a set of materials for each task, including a description of the mathematics in the task, scoring criteria on a scale from 0 to 4, approximately fifteen scored responses from students, and rationales for the scores assigned to each response. The tasks and materials are appropriate for use in the upper elementary and middle school grades, or approximately grades 4 through 8. Further, detailed descriptions of six activities illustrating various uses of the tasks and materials are included. A complete description of the contents of this book appears in chapter 1 in the section "Description of Tasks, Materials, and Activities."

The Tasks: Origin, Selection, Description, and Mathematical Content

THIS chapter presents an overview of the sixteen tasks used in this publication, including their origin and selection, their descriptions, and their mathematical content.

Where Did These Mathematics Tasks Come From?

The tasks included in this book and CD are a subset of those on the QUASAR Cognitive Assessment Instrument (QCAI) (Lane 1993; Lane et al. 1995). The QCAI is a mathematics performance assessment that was designed to measure student outcomes and growth in mathematics. It was administered over a period of five years, from 1990 to 1995, to all sixth-, seventh-, and eighth-grade students in six middle schools across the nation that participated in the QUASAR project[1] (Silver and Stein 1996). QUASAR (Quantitative Understanding: Amplifying Student Achievement and Reasoning), initiated in 1989, was a national mathematics education project with a reform orientation. The aim of the project was to demonstrate both the feasibility and responsibility of assisting middle schools in economically disadvantaged communities in the design and implementation of high-level, meaning-oriented instructional programs emphasizing mathematical thinking, reasoning, and problem solving.

Many of the tasks on the QCAI were developed soon after the first version of *Curriculum and Evaluation Standards for School Mathematics* (NCTM 1989) was released. The tasks were created to align with fundamental features of the reform-oriented conceptualization of mathematical proficiency. The format of the tasks is similar to that of the extended constructed-response tasks that appear on the mathematics portion of the National Assessment of Educational Progress (Silver, Alacaci, and Stylianou 2000) in that students are required to produce responses that include explanations, justifications, descriptions of solution processes, and diagrams. Each task takes approximately five minutes to solve.

Consistent with the views of mathematics instruction and assessment in *Principles and Standards for School Mathematics* (NCTM 2000), the QCAI covered a broad range of mathematical content areas, going beyond numbers and operations to include estimation, measurement, geometry, data analysis, probability, statistics, patterns, and functions. The tasks also assessed a variety of cognitive processes, such as understanding and representing problems, discerning mathematical relationships, organizing information, using and discovering various solution strategies, formulating conjectures, evaluating the reasonableness of answers, generalizing results, and justifying solutions.

Responses to each task were scored on a scale from 0 to 4 using a focused holistic rubric. Middle school mathematics teachers were trained in formal sessions each summer to rate students' responses. The QCAI general rubric incorporated three interrelated components that reflected the conceptual framework of the assessment: (1) conceptual and

---

[1] The QUASAR project was funded by the Ford Foundation. Six middle schools in economically disadvantaged communities across the country participated in the project from the 1990–91 school year to the 1994–95 school year. The ethnicity of the participating student population was diverse, encompassing African Americans, Latinos, and Caucasians and a smaller percentage of Asian Americans and Native Americans.

procedural knowledge of mathematics, (2) strategic knowledge of mathematics, and (3) mathematical communication (see appendix A). Specific rubrics were developed for each task to ensure high interrater reliability. Shortened versions of each specific rubric are included for all sixteen tasks in this book.

During its development, the QCAI went through a series of rigorous internal and external reviews. All tasks were evaluated by several professionals in mathematics education to ensure that they measured important and meaningful mathematical concepts and processes, that the wording was at the appropriate level for the students, that the tasks were interpreted in the same manner by all students, and that the tasks were free of bias and the possibility of different interpretations stemming from differing cultural backgrounds and experiences. Reviewers also examined each specific rubric for consistency with the general rubric criteria and to ensure that appropriate distinctions were made among the levels of mathematical understanding for each task.

After the review, tasks were pilot tested before placement on the assessment. Students' responses on the pilot test were analyzed to ensure that the tasks worked as intended. For example, did all students interpret the task in the same manner? Did the tasks evoke cognitively complex performances? Did the tasks elicit a variety of solution strategies? On the basis of the outcomes of the pilot tests, tasks were modified if necessary and retested. During the years that the QCAI was administered in the schools, several studies were conducted to ascertain the validity and reliability of the assessment (see, e.g., Lane and Parke [1995]; Lane and Silver [1995]; Lane et al. [1994]; Magone et al. [1994]).

How Were the Tasks Selected?

THE COMPILATION of the tasks and materials in this book was guided in part by what was known about the functioning of the tasks on the assessment instrument. Other sources of guidance included information on how the QCAI tasks, QCAI rubrics, and students' responses were used over the years by middle school teachers; mathematics educators, called resource partners, who worked with teachers at each school; and others in the QUASAR project.

Site-based teams at each school were provided with results indicating the level of students' performance on all tasks, as well as in-depth information about students' performance on a subset of tasks. This information included the nature of students' mathematical

understanding and communication, solution strategies and representations used to arrive at the answers, and common misconceptions and errors. For the released tasks, teachers were also given sample sets of scored responses and the entire set of actual responses from their schools. The teams were encouraged to use the materials in classrooms with students and in sessions with groups of teachers.

We learned about the many ways the tasks and materials were useful to the schools as we made school visits, conducted classroom observations and workshops, and held discussions with teachers and resource partners (see, e.g., Parke and Lane [1997]). At one point after the project's completion and during the creation of this book, a group of QUASAR teachers was convened to give their thoughts about which QCAI tasks should appear in the book. General recommendations were that the included tasks should cover a broad range of important mathematics; focus on conceptual understanding; permit a variety of solution strategies; and allow for the use of diagrams, sketches, and other "thinking tools" in the justification of the solution. Because of the nature of the assessment, these qualities described nearly all the QCAI tasks.

Guided by these suggestions and what we learned over the years about the variety of instructional and assessment settings in which the tasks were used, we have transformed the original assessment instrument into a resource for mathematics teachers and other education practitioners.

Description of Tasks, Materials, and Activities

THE SIXTEEN tasks in this book cover a range of mathematical content areas. Although some tasks address more than one content area, the primary mathematical topic is identified below for each of them. Number and operations are addressed in four tasks; measurement and geometry, in five tasks; data analysis and probability, in four tasks; and algebra, functions, and patterns, in three tasks. Descriptions of the specific mathematics in each task are found in the section titled "Mathematical Content of the Sixteen Tasks." A "task packet" has been created for each task, consisting of—

- a description of the mathematics in the task and the possible solutions;

- a blank copy of the task;

- scoring criteria for each level from 0 to 4;

- approximately twelve to fifteen scored responses, that is, two to three responses representing each score level; and

- a rationale for each scored response, explaining how the criteria were used to assign a score.

All sixteen task packets are included on the CD for ease of printing and reproduction. Five of the sixteen task packets are also reproduced in the book in the chapter titled "Task Packets." These five tasks are used as examples in the activities.

Listed below are the six activities that describe ways in which the tasks and materials can be used. In activities 1 and 2, students examine responses to mathematics tasks, explore various ways to approach complex problems, and discuss misconceptions and errors in responses. Activities 3 and 4 familiarize students with differing levels of mathematical understanding by asking them to apply established criteria to score responses or to develop their own criteria to compare and evaluate responses. Finally, activities 5 and 6 describe how teachers can use the tasks to determine students' existing knowledge and assess students' knowledge for the purposes of making informed decisions about instruction. The activities are as follows:

Activity 1: Exploring Multiple Strategies, Representations, and Answers

Activity 2: Editing Responses to Improve Their Quality

Activity 3: Using Established Scoring Criteria to Evaluate the Quality of Responses

Activity 4: Developing Scoring Criteria to Evaluate the Quality of Responses

Activity 5: Assessing Students' Existing Knowledge

Activity 6: Monitoring Students' Learning during Instruction

The intent of these activities is to illustrate the potential value to both teachers and students of using performance tasks in the classroom. Although each activity is carried out using only one of the sixteen tasks, other tasks included in this book would work just as well. Suggested tasks that are especially suited for each particular purpose are included in the description of each activity. No sequence is suggested for using the activities; however, the first four involve students' exploring solutions and explanations of mathematical thinking, and the last two describe teachers' using the tasks to guide instruction. Depending on the level of familiarity with performance tasks, teachers can select those activities and tasks that will best meet their students' needs and enhance their own instructional programs. Both students and teachers will benefit from using the tasks and students' work. The activities encourage students to become reflective thinkers and problem solvers through examining and discussing sample responses. In addition, when several teachers in the same school use the activities, valuable teacher-to-teacher conversations can be initiated about varying levels of students' performance and about the decisions teachers make in their classroom instruction.

Note that the QCAI general rubric (see appendix A) was used to assign scores to the samples of students' work in each task packet in this book, and the list of rationales for each scored response reflects the criteria described at each level of performance in the rubric. The holistic five-point rubric used with the QCAI incorporates criteria to assess conceptual and procedural knowledge in mathematics, strategic knowledge in mathematics, and mathematical communication and assigns one overall score to each student's response. The intention of using this rubric to score students' responses was simply to aid in illustrating the various ways performance tasks can be used and to give readers a starting point for using the task packets. A teacher may want to use a familiar rubric, such as one used in the local school, the district, or the state, or may want to develop an individualized rubric. A rubric with four or six levels may be preferred over five levels. A teacher might want to assign one score to the accuracy of students' procedures and calculations, another score to the level of mathematical understanding exhibited, and a third score to mathematical communication, thus using a more analytic approach to scoring rather than a holistic one.

The important issue is not the exact score that is assigned to a response but the discussions that take place in getting to that score. For example, the teacher and student conversations in activity 1 revolve around the nature of the solution strategy used by students and how completely and correctly the strategy is applied rather than focus on the exact assigned numerical score. Similarly, in activity 2, students revise lower-level responses to improve them, thus creating higher-quality responses. Another activity explicitly asks students to evaluate the quality of responses using their own judgments. Activity 4 describes a classroom situation in which the assigned scores have been removed from all responses. Students sort the responses into high-, mid-, and low-level categories, then develop their own rubrics by creating scoring criteria for each level.

We hope that the ideas and suggestions in this book will spark your interest in using the tasks and scoring materials in other ways. You may want to create extensions for some tasks or incorporate the use of manipulatives or collaborative group work. Furthermore, the first four activities, which illustrate classroom uses of the materials, can easily be used with teachers in professional development sessions, as discussed in the section titled "Teachers' Professional Growth and Development."

Mathematical Content of the Sixteen Tasks

THE FOLLOWING paragraphs describe the mathematics in each task. The sixteen tasks are listed by their primary content-area focus, including numbers and operations; measurement and geometry; data analysis and probability; and algebra, functions, and patterns. Asterisks indicate tasks that are used to illustrate the sample activities.

- Blocks*—Numbers and Operations

  This task allows students to demonstrate their understanding of number sense and their problem-solving abilities by using basic concepts of number theory. Students are asked to find the total number of blocks that when placed in groups of two, three, and four result in one block remaining. Two constraints must be satisfied: (1) the same total number of blocks is partitioned each time into groups of two, three, and four; and (2) the total number of blocks has a remainder of 1 when divided by 2, 3, and 4. Students are asked to provide an answer and show their work. A number of strategies can be used to solve the problem, such as finding common multiples, interpreting remainders in division computations, and using diagrams to represent groups of blocks. Multiple correct answers are also possible for this task, including 13, 25, 37, and so on.

- Margarita and Sam—Numbers and Operations

  In this task, students must compare rates of money earned per day by two children. They

are asked to determine the number of days each child could work that would result in the same total amount of money earned by each child. Students are asked to find two sets of answers and must show how they obtained these answers. Many correct answers are possible for this task, such as the following:

(2, 3)   (4, 6)   (6, 9)   (8, 12)

(10, 15)   (12, 18)   (14, 21)   (16, 24)

(18, 27)   (20, 30)   (1, 1$\frac{1}{2}$)   (3, 4$\frac{1}{2}$)

(5, 7$\frac{1}{2}$)   (7, 10$\frac{1}{2}$)   (1$\frac{1}{3}$, 2)

- Mathematical Discovery—Numbers and Operations

  The purpose of this task is to determine students' understanding of the effects of multiplication on whole numbers, fractions, and decimals. Students are given an incorrect conjecture stating, "When you multiply *any* two numbers, the answer will always be larger than *each* of the two numbers you multiply." Students are asked whether they agree with the statement. A correct response to this task includes the development of a mathematical argument using one or more counterexamples to disprove the statement.

- Triangles—Numbers and Operations

  This task requires students to demonstrate understanding of part-to-whole relationships. Students must interpret and modify a pictorial representation of a discrete set of black and white triangles to achieve a given numerical fraction. Specifically, students are asked to add black triangles so that half of all triangles are black, meaning that one black triangle needs to be added to the set. They are also asked to remove black triangles so that one-third of the remaining triangles are black, meaning that two black triangles must be removed from the set. For both questions, students' explanations must refer to the comparison of the number of white and black triangles with the total number of triangles.

- School Board*—Measurement and Geometry

  The purpose of this task is to determine whether students can differentiate between the concepts of area and perimeter and use them appropriately in two problem situations. The terms *area* and *perimeter* do not appear in the task. Instead, the task asks students to demonstrate that they can apply the appropriate measurement when given particular situations. Specifically, students are shown figures of two rectangular plots of land with different dimensions, and they must identify which figure has the larger area, that is, would enclose "as much land as possible," and which figure has the smaller perimeter, that is, would require "less fencing." Students are also asked to show how they found their answers; doing so usually includes showing calculations and comparisons of areas and perimeters of the figures.

• Double the Carpet*—Measurement and Geometry

This task requires students to determine and compare the areas of two figures. The dimensions of two rooms are provided in a problem situation. The perimeter of the game room is double that of the living room. Students are asked to determine the correctness of the assertion that the "area to be carpeted in the game room is double the area to be carpeted in the living room." Students must describe how they made this decision. Arithmetic calculations, written explanations, and diagrams may be used to show that the area of the game room is not double the area of the living room but, rather, four times as large.

• Art Class—Measurement and Geometry
In this task, students are asked to perform a simple tessellation. Students must find the maximum number of given identical shapes, in this instance, three-by-five-inch cards, that will cover a finite region, a 12-by15-inch cardboard sheet, without overlapping. The answer is 12. Students' work, diagrams, or explanations for this task should contain the following two features: (1) correct orientation of small cards on a rectangular cardboard sheet, and (2) complete, nonoverlapping coverage of the cardboard sheet. A variety of representations can be used to show these two features.

• Art Project—Measurement and Geometry

This task requires students to find the area of an irregular six-sided shape that is composed of two smaller overlapping shapes, one of which is a square and the other, a triangle. Diagrams of the two original shapes and the one larger shape are shown, and some dimensions are given. Using the diagrams and dimensions, students must find the area of the irregular larger shape and explain how they found the answer. The answer is 88 square centimeters. A variety of solution strategies can be used, including the following:

- Treating the irregular shape as an overlapping square and triangle, then finding the combined area of the square and triangle and removing the area of overlap

- Treating the irregular shape as a large rectangle with a missing triangle, then calculating the area of the large rectangle and removing the area of the small triangle

- Treating the irregular shape as the composition of several small regular figures, then calculating the area of the subdivisions and combining to obtain the area of the irregular shape

• Mr. Jackson's Fence—Measurement and Geometry

This task allows students to demonstrate an understanding of the concept of perimeter of a rectangle; however, the term *perimeter* does not appear in the task. Instead, a specific amount of fencing is given, and students are asked to determine the lengths and widths of two possible rectangular spaces that would use all the fencing and show how they found their answers. Several correct answers are possible for this task. Parts A and B should each have a unique and correct length and width and should be accompanied by work showing that the sum of the sides is 50 feet.

• Left-Handed Survey—Data Analysis and Probability

This task requires students to interpret and manipulate data from a sample, then compare a given percent with the percent obtained in the sample. Four subsets of the total sample are presented in tables. Students are asked to determine the total percent of observations of a certain type across the subsets and to compare that result with a benchmark of 10 percent. Students must show or explain the basis for their comparisons. Several strategies can be used to determine that the overall percent derived from the four classrooms is larger than the benchmark percent. Some of these strategies are outlined in the following:

- Find the percent of left-handed students in the survey by dividing the number of left-handed students, 9, by the total number of students, 80, to obtain 11.25 percent, which is larger than 10 percent.

- State that 10 percent of the total number of students, 80, is 8 but that the survey found 9 left-handed students in the four homerooms; therefore, the percent of left-handed students in these homerooms is larger than 10 percent.

- State that 10 percent of the number of students in each classroom, 20, is 2, then compare the number of left-handed students in each room with 2 and use a "leveling off" strategy to determine that the overall percent is larger than 10 percent.

- Find the exact percents for each homeroom, then average these percents to arrive at an overall answer that is larger than 10 percent.

• Bar Average*—Data Analysis and Probability

This task focuses on students' understanding of the concept of average. The set of data, represented in a bar graph, consists of scores on three 20-point projects. Students are asked to find the score on the fourth 20-point project when given the average of all four project scores and to explain or show how they found the answer. This task allows for a variety of solution strategies, such as the following:

- Applying the concept of average to determine the total number of points needed across all projects, then working backward to find the answer; for example, $17 \times 4 = 68$, $68 - (15 + 18 + 16) = 19$.

- Substituting x in the average formula for the fourth project, then solving the algebraic equation; that is, $(15 + 18 + 16 + x) \div 4 = 17$, therefore $x = 19$.

- Applying a "leveling" strategy, for example, moving points from one project to another

to maintain 17 points for each of the four projects. Students can then determine that the score for the fourth project must be 19 points.

- Guessing scores for the fourth project, then checking each score using the formula for average until the average of 17 is obtained; for example, $15 + 18 + 16 + 19 = 68$, and $68 \div 4 = 17$.

• Jamal's Survey—Data Analysis and Probability

This task describes a problem situation in which a survey was administered to a sample obtained from a population. Given information about the sample and the population, students are asked to provide a result of the sample data, then make inferences about the whole population. Specifically, in a population of 320 sixth graders, 32 students were surveyed. One-half of the students surveyed said that they listen to music after school. In part A, students are asked how many students in the survey sample listen to music; the correct answer is 16. In part B, students are asked to use the results of the survey to predict how many students in the whole population listen to music; the correct answer is 160. The response can include calculations, written explanations, or pictorial representations of the sample and population. Part B should include a reference to the survey's results.

• Spinner—Data Analysis and Probability

This task allows students to demonstrate basic understanding of probability. Students determine the probable number of occurrences of certain events when a spinner is used. The spinner contains four fractional pieces: two are white, one is black, and one is gray. When given a total number of spins, sixty, students are asked to determine how many times they would expect the arrow to land on white and how many times they would expect the arrow to land on black. They are also asked to explain their answers.

For the answers to parts B and C, close estimates other than the theoretical values of 30 and 15 can be allowed. The following are characteristics of a "good" prediction:

- If predictions are whole numbers, they should be based on sixty spins, and part B should be about twice as large as part C. For example, predictions of 30 in part B and 15 in part C would be reasonable, as would predictions of 34 in part B and 14 in part C.

- If predictions are ranges of numbers, the range in part B should contain 30 and the range in part C should contain 15. For example, a reasonable range might be 25 to 35 in part B and 10 to 20 in part C.

- Miguel's Homework*—Algebra, Functions, and Patterns

  In this pattern task, students demonstrate the ability to recognize the underlying mathematical structure used to generate a visual pattern, then describe the pattern. Specifically, students are asked to draw the fifth figure in a pattern, given the first four figures. Students must also describe how they know which figure comes next in the pattern. This visual pattern involves two regularities: (1) the number of dots added to each figure, and (2) the shape of the figure or the positions in the pattern where the dots are to be added. The students' drawings and descriptions of the pattern must include both of these regular features; that is, (1) three dots must be added to each subsequent figure, and (2) one dot must be added to each row of the figure to get the next figure.

- Tony's Walk—Algebra, Functions, and Patterns

  This task involves students in reading, interpreting, and integrating information given in a graph and using that information to write a short story. Students are given a line graph that shows a relationship between speed and time and are asked to write a story in which events or physical phenomena would correspond to the relationships between speed and time. Four speed concepts are represented in the graph and should be incorporated in the stories that students write: (1) moving at a constant slower pace, (2) moving at a constant faster pace, (3) slowing down, and (4) stopping.

  Students might account for Tony's time and speed in several ways in their stories. For

example, the concepts listed above might be explicitly stated. A student's understanding of these speeds might also be implicitly expressed through a description of such activities as walking, resting, getting tired, and running or through an estimate of speeds at various times, such as walking at 3 miles per hour from noon until 12:30 P.M.

- Brenda's Blocks—Algebra, Functions, and Patterns

  In this task, students are given written and pictorial information that represents two interrelated equations. One diagram of a balanced scale shows that three plastic blocks weigh the same as two wood blocks. The second diagram shows that one plastic block and one wood block weigh a total of 15 ounces. Students are asked to determine the weight of the two unknown quantities, that is, the plastic block and the wood block. The correct answers are 6 ounces for one plastic block and 9 ounces for one wood block. Different solution strategies can be used to solve the problem, including the following:

  - Set up an algebraic equation. If x is the weight of one plastic block, then $3x = 2(15 - x)$, and $x = 6$; therefore, one plastic block weighs 6 ounces and one wood block weighs 9 ounces.

  - Use the ratio of a wood block to a plastic block, 3 to 2. Together, the ratio has 5 units. These 5 units are equal to 15 ounces, and each unit is 3 ounces; therefore, a plastic block is 3×2, or 6, ounces and a wood block is 3×3, or 9, ounces.

  - Use a guess-and-check strategy. Choose pairs of numbers and check them using the two interrelated equations represented in the problem until the answer is obtained.

Chapter 2

Teacher Commentary on Using Performance Tasks and Rubrics

TEACHERS IN the QUASAR project used the QCAI tasks and scoring materials in a variety of ways. During the years in which the QCAI was administered, teachers used the released tasks. On completion of the project, the entire assessment instrument was released, thus allowing teachers access to all tasks. Six detailed examples of the uses of the tasks and materials are presented in this book. This section contains a more general discussion compiled from many interactions with teachers regarding the benefits of using mathematics performance tasks and scoring rubrics.

Students and Teachers Exploring Multiple Solution Strategies and Representations

MANY TEACHERS expressed the idea that one major advantage of performance-based tasks is that they allow students to solve problems in more than one way. In the classroom, having students share the paths by which they arrived at their answers helps others realize that many approaches can be taken in solving a problem. Instead of talking only about the correctness of the answer, teachers and students had deeper discussions about the mathematics in the problem. As one teacher said, "My students could see how two very different responses can both be correct." When looking at the set of anonymous responses to tasks, teachers and students also talked about the appropriateness of certain solution strategies for certain problems and whether the solution strategies were implemented correctly and completely. Teachers might ask such questions as "Would this solution strategy lead to the correct answer in this problem?" or "This student used an appropriate strategy, buy why did he get an incorrect

answer?" Activity 1 in this book uses sample responses from the Blocks task to engage students in exploring various methods of arriving at a solution. This task also has more than one correct numerical answer, an eventuality that makes for an even richer classroom discussion.

Another teacher said that her sixth- and seventh-grade students came up with five different ways of solving the Art Project task; this teacher commented, "The way[s] they divided the figure were interesting." (The complete packet for the Art Project task, including samples of students' work, is found on the CD accompanying this book.) In using such tasks, teachers often learned a lot about their students; they were pleasantly surprised by students' unique solution strategies that even they themselves had not considered. Other teachers also talked with students about the different types of representations that can be used to solve a problem. While working on a unit involving patterns and functions, one teacher made this comment:

> What's important is that kids see that there's more than one way to deal with patterns and functions. They can describe a pattern by drawing pictures, writing the pattern in words, constructing a table of possible solutions, generating a graph, or writing formulas and algebraic equations.

Ultimately, the benefit of exposing students to different solution strategies is that when students approach a novel problem, they will have ready access to a number of different ways to solve it. One teacher aptly stated, "By giving students more opportunities to see how others approach problems and to practice higher levels of thinking, they will become better problem solvers."

Students and Teachers Discussing Mathematical Communication

MATHEMATICS TASKS that ask students to explain how they arrived at their answers or to justify their solutions help them develop their mathematical communication skills. Students' first exposure to writing in mathematics, however, is often a daunting one. One teacher made this observation:

> When I first asked students to explain their answers in writing, their responses were trite and lacked depth and clarity. But the more I incorporated tasks like these and we talked about what a complete answer looked like, they gradually learned how to put their mathematical thoughts into words. It helps to give students an opportunity to show their thinking on tasks like this throughout the year so that [writing] becomes part of your regular classroom activities.

As students begin to put their thoughts into words, whether on paper or orally, they need to see examples of effective mathematical communication. Teachers often noticed that some written explanations were quite lengthy but did not focus on the mathematics in the problem. One teacher said, "It's not that they don't write, but it's what they write." Another teacher remarked, "There are still some students who think that if they write something down and turn it in, then it's okay. But I consider that to be a 'work in progress.' In the end, it's not okay; it needs to make sense mathematically. Thirty pages of nonsense isn't quality work." To aid students in writing better explanations, teachers compared samples of students' explanations that were complete and correct with samples that were not. One teacher remarked,

> We looked at one response that had a lot of writing but didn't really say anything and another response that had very little writing but was complete. I spend a great deal of time in my class talking about "quality writing," and I tell my students [that] they shouldn't put down a lot of words just for the sake of doing so.

Teachers noted that another benefit of using complex, interesting, and worthwhile mathematics problems was the improved interactions among the students and teacher during whole-class discussion and in cooperative-group settings. One teacher commented, "I can see more student input in class discussions now. Students share more with others about how they are thinking about a problem. [Communicating their thinking] increased students' participation, and they enjoy math more because it is alive and active."

Activity 2 promotes this type of classroom discussion. First, students are shown samples of high-quality work. Then they examine responses that are incomplete, unclear, or incorrect and talk about how to improve these responses. This process helps students get a better sense of what is expected in a mathematical explanation and encourages students' self-assessment of their work. Activities 3 and 4 also foster mathematical communication by giving students a chance to assign scores to other students' work on the basis of the completeness of the responses.

Students and Teachers Using Rubrics

SHARING RUBRICS in the classroom helps students gain a better understanding of the elements of a good, complete response to a problem. Examining high-quality work gives students a sense of the standards they are expected to meet. One teacher said, "When I show them my rubric, I am able to communicate clearly with students how they will be evaluated, and we all know exactly what to look for as far as examples of understanding."

Sharing rubric criteria and scored responses for a task can be helpful for students who are just beginning to familiarize themselves with tasks that require higher-level thinking and reasoning. After discussing scoring criteria with her students and looking at high- and low-level responses, a teacher remarked, "My students saw in more detail what a really good explanation looks like. Seeing an exemplary paper was very good for them; it helped students know what is expected. They also saw what [responses getting scores of] 0's and 1's looked like." Teachers sometimes talked about the elements that were omitted from the lower-level responses or those that were incorrect, then had students edit these responses to improve them.

When the students applied scoring criteria to evaluate a new set of responses, teachers were pleased to see that they usually graded their own and others' work fairly. A teacher commented that his students "really enjoy this activity, and it engages them in meaningful interactions." Teachers also gain insight into students' understanding and the features of a response that students believe are important. One teacher noticed the following about her students' perceptions:

> I observed some students giving higher scores to papers that had long explanations. They were valuing someone who would write a lot even if their explanation was way off. This [misconception] is something I need to address in my instruction.

This scenario is similar to activity 3, in which students become familiar with an established set of criteria, then apply it to evaluate a set of responses. After students score the responses, the scores are tallied and students identify and discuss those responses on which they reached a high level of agreement in scoring. Students also identify, and have more lively discussions about, responses on which they disagreed in scoring. Teachers found that these types of activities encouraged more intense self-reflection when students are asked to produce their own work for performance tasks.

In other classrooms, teachers had their students look at unscored responses, then develop their own rubrics. Activity 4 in this book describes a similar classroom situation. Either in a whole-class setting or in groups, students sorted responses into different categories, for example, high, midlevel, and low, on the basis of the quality of the responses. After coming to an agreement about the placement of responses and talking about why they put a response in a particular category, students created their own scoring criteria. Teachers enjoyed listening to students' animated interactions about how to score responses and how to write the criteria. As one teacher said, "I was glad that students were having such good conversations about the scores. This activity encourages them to really think about the math in the problem, and ultimately, it deepens their reflections on their own work."

A related issue is the fact that different rubrics have different expectations for achieving the highest-level response. This idea is important to discuss with students when they are attempting to develop their own rubrics. For example, can a response be classified at the highest level if it answers the task completely and correctly, or does the work need to go above and beyond what is asked in the task to achieve the highest level? Is providing a mathematical explanation in words and sentences always necessary for every task? Is a clearly labeled, annotated drawing or sketch good enough? Should the response provide more than one correct answer? (See activity 2 for a discussion of a task with multiple correct answers.) What score is given to a response that shows the correct answer but without supporting work or explanation? Ultimately, any group involved in developing a rubric, whether students or teachers, must eventually grapple with, and reach consensus on, such decisions.

Teachers Improving and Modifying Instruction

As one teacher said, "Asking my students to solve these tasks and looking at their work [have] made me more aware of what's going on in my class. It gives me

a more accurate perspective of what students do and do not understand about a topic." Many teachers made similar comments regarding how they use information from performance tasks when planning instruction. Misconceptions that students may have or confusion that they may experience about a particular concept are often brought to light, allowing teachers to adjust the lesson to address these issues.

After one teacher examined a set of student responses that all had the correct answer but displayed different kinds of thinking about the problem, she made this comment:

> To me as a teacher, there's a lot of value in doing this. I would like to know what they don't understand. I read these responses and say, "They may be getting the answer right, but I'm real[ly] uncomfortable with the way they're talking about it."

This knowledge about their students can help teachers make decisions about the way content is presented. Put simply by one teacher, "[Analyzing students' responses] opens our eyes as to what was taught versus what was learned."

Performance-based tasks can both guide instruction and also give teachers evidence for making informed judgments about students' knowledge and understanding. One teacher remarked, "Looking at the data from performance-based items is rich. It gives us information that we just don't get from other sources." In fact, these data help to complete the picture of a student. Another teacher said, "I can see how my students are thinking and the problem-solving processes they are using. I can better determine if students have grasped the concept, as opposed to just getting the answer right."

The sections on activities 5 and 6 in this book present more specific examples of how teachers used students' work from performance tasks to glean information about how their students think and to make instructional decisions stemming from the levels of students' performance. The tasks used for both these activities assess students' knowledge of area and perimeter. In the section on activity 5, a teacher uses the tasks to assess students' knowledge before instruction. In the section on activity 6, a teacher uses the tasks near the end of an instructional unit to determine the levels of students' knowledge about area and perimeter and to make decisions about what concepts should be revisited in instruction.

Teachers' Professional Growth and Development

THE TASKS and related scoring materials in this book can also be used as the basis for professional development experiences for mathematics teachers. Because the tasks and scoring guides embody important ideas found in the middle-grades mathematics curriculum and because the responses illustrate the kinds of thinking that might typically be found among students at this age, the material in this book can be a resource for what has become known as "practice-based" professional development. In her seminal book on the notion of practice-based professional development for mathematics teachers, Smith (2001) illustrates ways in which "samples of authentic practice—materials taken from real classrooms—[can] become the curriculum for teacher education by providing opportunities for critique, inquiry, and investigation" (p. 7). In fact, this kind of experience was indispensable in the QUASAR project.

At the beginning of the QUASAR project in 1990, many teachers were just beginning to encounter complex tasks that challenged students to think, reason, and explain. Sample QCAI tasks and sample scored responses were often used during early workshops with teachers at participating QUASAR schools to acquaint them with the demands that such tasks would place on their students and with the rich information that performance tasks could yield regarding their students' thinking and learning. (Mathematics educators who examined students' responses to NAEP tasks noted the utility of such tasks for revealing important facets of students' thinking; see, e.g., Kenney, Zawojewski, and Silver [1998]; Stylianou et al. [2000]).

At some QUASAR schools, professional development experiences sometimes took on a form similar to that of activities 1 and 2 in this book, which are written, here, for use with students in the classroom but also functioned well when used with groups of teachers during professional development sessions. For instance, QUASAR teachers at most schools administered the sample QCAI tasks, or released tasks in subsequent years, in their classrooms and brought students' responses to teacher meetings. During the meetings, they examined the work produced by the students and analyzed the kinds of thinking, problem-solving strategies, and errors revealed in the responses. They also read responses with varying explanations and compared the levels of understanding revealed. Typically, a few of the explanations would be quite good and others would be vague or incorrect. At first, a few teachers in the school could not understand the necessity of "forcing" students to explain in words when they had obtained the correct answer. After looking at task responses, however, these teachers were surprised to discover that even though students arrived at a correct answer, some of them did not have a good understanding of the concept. They began to appreciate various facets of students' understanding and mathematical conceptualization that are present in the responses to open-ended tasks. At one school, a teacher remarked,

> I'm starting to look at the way that students think, work, and solve a problem as the most important part of some assignments. I can know more about my [students]. I used to settle for just an answer, and now, they have to explain, and I can understand more about what is going on in their heads.

Once the teachers at QUASAR schools felt comfortable with tasks that required complex responses, the resource partners at the project sites began to focus on scoring criteria and rubrics. Groups of teachers engaged in activities similar to this book's activities 3 and 4, which are intended, here, for use with students. Teacher leaders often commented that the "sorting activities," which is how they referred to these assessment activities, familiarized teachers with scoring rubrics, enabled them to have productive conversations about issues related to scoring, and increased their appreciation of the value of scoring students' work in this way. One teacher said, "Using a rubric helps me simplify things and focus on key mathematical concepts in the problem rather than [get] caught up in looking only at poor computation skills. It's a good reminder that there are other ways to assess." After looking at assessment data from her class, another teacher commented that evaluating students' responses in relation to the scoring levels in a rubric "opens my eyes to the growth that is possible for our students. The way they enter a problem and with what types of knowledge [are] amazing to me."

As stated in *Principles and Standards for School Mathematics* (NCTM 2000), "much of teachers' best learning occurs when they examine their teaching practices with colleagues" (p. 370). One way to achieve this learning is by analyzing and discussing teaching episodes presented on videotape or in narrative cases (see, e.g., Stein et al. [2000]). Professional development sessions that focus on innovative curriculum materials can provide a platform for collegial inquiry into teaching. Lesson study is yet another vehicle to spark valuable discussion among professionals. For example, mathematics educators who use the Japanese lesson-study approach (see, e.g., Curcio [2002]) find that when novice and expert teachers take an active role in their professional development by "analyzing mathematics content, discussing instructional strategies, and anticipating students' responses to tasks and activities" (p. 7), they strengthen their content and pedagogical knowledge.

Each activity in this book gives teachers a chance to examine the responses produced by others' students or to watch other teachers in the school try the materials in the classroom. Colleagues then talk about the implementation of the materials, as they might in a case discussion, curriculum implementation workshop, or lesson-study session. Discussing students' work often served as a springboard for meaningful interactions among the QUASAR teachers about their teaching practices. Examining the thought processes evident in their students' responses to tasks in the classroom and on paper often encouraged collaboration among teachers beyond the whole-group, in-service sessions. In one instance, a teacher gave a task to her students, scored the responses, then met with another teacher. The second teacher independently scored the responses, and the two colleagues examined the extent to which their scores agreed. One of the many issues they discussed was the importance of consistently applying the criteria in a scoring rubric and "not allowing what we know about the student from class to influence the evaluation of what is on the piece of paper."

Finally, the site-based teams at some of the QUASAR schools used the QCAI tasks and rubrics as models to develop their own tasks and rubrics. After administering these tasks in their classes, teachers met to discuss how the criteria in the QCAI general rubric could be applied to measure students' knowledge of the specific mathematics in each task. They discussed the rubrics and responses until they reached a shared definition of each of the scores. In these sessions, teachers had the opportunity to reflect on the mathematical concepts they had presented and the ways in which students demonstrated their grasp of these concepts. Teachers commented that this experience was quite valuable. In fact, one teacher said, "These were some of the best discussions we ever had about what the students know, what they are learning, and what evidence is on this piece of paper that shows what they know."

At another school, teachers developed their own rubrics to reflect some aspects of students' responses that they thought were not examined completely on the QUASAR rubrics. One teacher at this school commented, "Although our rubrics are not identical, looking at QCAI rubrics gave us a starting point and reinforced what we were doing." The work involved in creating rubrics was much more difficult than the teachers had anticipated, but they all agreed that the exercise was a good learning experience. Throughout the process, teachers debated several fundamental issues. First, they had to agree on the purpose for the rubric. Was the rubric only for teachers to use, or was it to be given to students? If it was a rubric for students, the writers had to make sure that the wording allowed students to understand the expectations without giving away too many clues that would help them solve the problem. Statements that were too general were almost useless to students, whereas other statements were too directive, thus defeating the purpose of open-ended items. Considerable time was spent making sure that the language in the rubric was meaningful to students. Another major obstacle was reaching agreement on the criteria for the highest scoring level. Should students be required to extend and generalize the task to receive the highest score? Should students be required to show a sketch or diagram in their explanations? Must a written explanation be supported or justified with numbers? Most of the decisions were made at the task level, thereby reflecting the nuances of individual tasks, but some general principles emerged. For example, evidence that students were pursuing generalization was considered to be essential in the distinction between performance levels at the high end of the scale.

Putting the Tasks to Work

THE FOLLOWING sections detail classroom activities that use the QCAI tasks to help students explore multiple solution strategies, encourage mathematical communication, and improve mathematical performance through an understanding of rubrics. Teachers may also use these activities to guide classroom instruction and foster their own professional growth. Of course, the activities may be adapted to incorporate other tasks from current instructional units; the important idea is to open the classroom to performance-based tasks and mathematical discourse to enhance the experience of students and teachers alike.

CHAPTER 3

Six Activities

Activity 1: Exploring Multiple Strategies, Representations, and Answers

PRINCIPLES and Standards for School Mathematics (NCTM 2000) identifies problem solving as an integral part of learning mathematics. When students are given "frequent opportunities to formulate, grapple with, and solve complex problems," they develop new mathematical knowledge and a variety of problem-solving strategies and approaches (p. 52). Many mathematics educators have argued that solving one problem in many ways is more beneficial for a student's development as a problem solver than solving many problems, each in only one way. Another important ingredient in the process of becoming a good problem solver is reflecting on one's strategies and thinking to learn from the experience. In activity 1, students are given opportunities to explore the use of multiple strategies and representations and to arrive at multiple answers to a task. In doing so, students reflect on their own solutions, as well as on those produced by others, in a way that helps them learn about their own problem solving.

Purpose of the activity

This activity is intended to engage students in exploring various methods of solving a task. Students are first asked to share their own approaches to finding solutions, then to discuss other students' approaches. Class discussions revolve around the many solution strategies and representations that can be used and the multiple answers to the task. Reflecting on their own approaches to the task and those of others helps students become better problem solvers by increasing their awareness of a variety of problem-solving strategies and by leading to

their realization that some tasks can have more than one correct answer.

Description of the mathematics task

The task chosen to illustrate this activity, called the Blocks task, involves number sense and basic concepts of number theory. Other tasks in this book and on the CD can also be used for activity 1. Art Class, Bar Average, Art Project, and Brenda's Blocks are a few tasks that have several interesting solution strategies. Like the Blocks task, Margarita and Sam is another task that has multiple answers.

In the Blocks task (see **fig. 3.1**), students must solve for an unknown number that satisfies several conditions set in a story context. Specifically, students must find the total number of blocks in a set given that one block remains when the whole set is partitioned into groups of two, three, or four. A condition of the problem is that the same total number of blocks is partitioned each time.

Implementation in the classroom

The set of sample responses in the Blocks packet contains several approaches to solving the task; however, before discussing these strategies with students, most teachers have students spend some time solving the task on their own to become familiar with the problem. After they have had a sufficient amount of time to think about the task and write their solutions, students can be asked to share their work with the rest of the class or in small groups.

Discussing solution strategies and representations

Students will likely present a variety of approaches to solving the task, but they may not generate all the interesting strategies and representations that teachers would like them to use. For this reason, teachers

Parke, Carol S., Suzanne Lane, Edward A. Silver, and Maria E. Magone. *Using Assessment to Improve Middle-Grades Mathematics Teaching and Learning: Suggested Activities Using QUASAR Tasks, Scoring Criteria, and Students' Work.* Reston, Va.: National Council of Teachers of Mathematics, 2003.

Blocks Task

Yolanda was telling her brother Damian about what she did in math class.

Yolanda said, "Damian, I used blocks in my math class today. When I grouped the blocks in groups of 2, I had 1 block left over. When I grouped the blocks in groups of 3, I had 1 block left over. When I grouped the blocks in groups of 4, I still had 1 block left over."

Damian asked, "How many blocks did you have?"

What was Yolanda's answer to her brother's question?

Show your work.

Answer: _____

Fig. 3.1. Blocks task

may want to have some of the sample responses in the packet available for the class to examine. The following examples illustrate the responses that Ms. Robertson shared with her students when she was using this released task in her classroom. The solution strategies included lists of common multiples, mathematical computations, and pictorial representations of blocks.

Example 1: Response A (see **fig. 3.2**) shows a list of all the multiples of 2, 3, and 4. Ms. Robertson talked with students about using the common multiple as a way to find the solution to this task. Twelve blocks can be grouped in twos, threes, or fours with no remainder. However, the problem says that one block was left over when the blocks were grouped. Thirteen blocks, then, will result in one block left over in each grouping.

Example 2: In response B (**fig. 3.3**), the student drew three separate diagrams in which the total number of blocks is a constant. Ms. Robertson's students examined the diagrams to determine what each of them represented. The first diagram shows blocks arranged in six groups of two blocks each. The second diagram shows four groups of three blocks each, and the third shows three groups of four blocks each. All three diagrams have a total of twelve blocks plus one remaining block. Ms. Robertson pointed out the other pictorial representations, which showed initial attempts to group blocks but were crossed out.

Example 3: Response C (**fig. 3.4**) shows mathematical computations that produce a common multiple for 2, 3, and 4, then 1 is added to represent the block that is left over.

Example 4: Response F (**fig. 3.6**) also shows a series of three calculations using multiplication, each having a product of 12. Then the division calculations are shown to verify that the answer of 13 would result in one block's being left over in each grouping. Because some of Ms. Robertson's students had difficulty interpreting this response, she talked with them about what features of the response are unclear. Multiplication is one way of finding a common multiple, but the solution does not clearly indicate how the answer of 13 was obtained. In fact, the way the division is represented is incorrect. The division should show 13 (blocks) divided by 12, not 12 by 13. At this point, if Ms. Robertson also wanted to discuss levels of responses, she could mention that this response was scored at a level lower than the previous responses. See activities 2, 3, and 4 for discussions of incomplete and incorrect responses.

After discussing the different solution types with her students, Ms. Robertson decided to extend the activity by asking them to classify a small set of students' responses according to strategy type. She used some of the remaining sample responses in the packet, as well as her own students' responses. Teachers might also use responses from another one of their own mathematics classes or from another teacher's class. Some teachers said that they remove names from the responses to eliminate the concern that some students might feel uncomfortable having their responses "evaluated."

Regardless of how responses are obtained, students can determine whether the approach to the task is similar to one of the strategies they discussed previously, a variation on the strategy, or a completely different strategy. If the strategy used is different, students can decide whether it is appropriate for the problem under consideration. Teachers who have done similar activities with other tasks reported that they had productive discussions in their classes about the merit of applying certain strategies to certain problems. Students discussed the idea that a strategy that is appropriate for one task may not necessarily be effective for another task. For those strategies that are deemed to be appropriate for the Blocks task, students can also be asked to determine whether the strategy was implemented correctly and completely. The section on activity 2, which also uses the Blocks task as an illustration, provides a more detailed example of students' examining errors and confusion in responses and editing the responses to improve the quality.

Discussing multiple answers

Another interesting feature of the Blocks task is that multiple answers are possible. In fact, the task has an infinite number of answers. If a teacher finds that none of the students comes up with an answer other than thirteen, she or he might want to have students solve the task again to find another total number of blocks that will meet the conditions. This task lends itself to a rich mathematical discussion in which students can create a more generalized solution. By examining lists of multiples, students can discover that the common multiples of 2, 3, and 4 are 12, 24, 36, 48, and so on. By adding 1 to each of these multiples, to represent the remainder of blocks after each grouping, a series of answers can be found. Response D (**fig. 3.5**) illustrates the work of a student who realized that many answers are possible. Note that this response was scored at the same level as responses A, B, and C.

Yolanda was telling her brother Damian about what she did in math class.

Yolanda said, "Damian, I used blocks in my math class today. When I grouped the blocks in groups of 2, I had 1 block left over. When I grouped the blocks in groups of 3, I had 1 block left over. And when I grouped the blocks in groups of 4, I still had 1 block left over."

Damian asked, "How many blocks did you have?"

What was Yolanda's answer to her brother's question?

Show your work. Well first I find a comon multiple of all three

2, 4, 6, 8, 10, (12), 14 16 18

3, 6, 9 (12) 15, 18

4, 8, (12), 16

12 is the first comon so I add 1 and my awnsers (13)

Answer: ___13___

Fig. 3.2. Activity 1—Blocks response A

B

4

Yolanda was telling her brother Damian about what she did in math class.

Yolanda said, "Damian, I used blocks in my math class today. When I grouped the blocks in groups of 2, I had 1 block left over. When I grouped the blocks in groups of 3, I had 1 block left over. And when I grouped the blocks in groups of 4, I still had 1 block left over."

Damian asked, "How many blocks did you have?"

What was Yolanda's answer to her brother's question?

Show your work.

Answer: _13 blocks_

Fig. 3.3. Activity 1—Blocks response B

C   4

Yolanda was telling her brother Damian about what she did in math class.

Yolanda said, "Damian, I used blocks in my math class today. When I grouped the blocks in groups of 2, I had 1 block left over. When I grouped the blocks in groups of 3, I had 1 block left over. And when I grouped the blocks in groups of 4, I still had 1 block left over."

Damian asked, "How many blocks did you have?"

What was Yolanda's answer to her brother's question?

Show your work.

$$4 \times 3 = 12 + 1 = 13$$
$$2 \times 6 = 12 + 1 = 13$$
$$3 \times 4 = 12 + 1 = 13$$

Answer: ___13___

Fig. 3.4. Activity 1—Blocks response C

D   4

Yolanda was telling her brother Damian about what she did in math class.

Yolanda said, "Damian, I used blocks in my math class today. When I grouped the blocks in groups of 2, I had 1 block left over. When I grouped the blocks in groups of 3, I had 1 block left over. And when I grouped the blocks in groups of 4, I still had 1 block left over."

Damian asked, "How many blocks did you have?"

What was Yolanda's answer to her brother's question?

Show your work.

2, 4, 6, 8, 10, (12), 14, 16, 18, 20, 22, (24), 26, 28, 30, 32, 34, (36)

3, 6, 9, (12), 15, 18, 21, (24), 27, 30, 33, (36)

4, 8, (12), 16, 20, (24), 28, 32, (36)

Answer: <u>13 blocks,</u> but there could be 25 blocks too, and it can go on like 12, 24, 36, 48, ···
and add 1 to each of them for the left over block to get
13, 25, 37, 49, ···

Fig. 3.5. Activity 1—Blocks response D

F   3

Yolanda was telling her brother Damian about what she did in math class.

Yolanda said, "Damian, I used blocks in my math class today. When I grouped the blocks in groups of 2, I had 1 block left over. When I grouped the blocks in groups of 3, I had 1 block left over. And when I grouped the blocks in groups of 4, I still had 1 block left over."

Damian asked, "How many blocks did you have?"

What was Yolanda's answer to her brother's question?

Show your work.

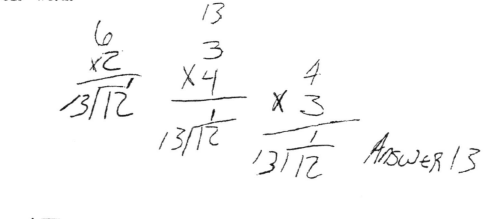

Answer: 13

Fig. 3.6. Activity 1—Blocks response F

If students are discussing score levels, then this response could receive a 4* to indicate that the response was above and beyond the requirements of the task.

One teacher commented, and others agreed, that "a major advantage of having students talk about responses to good, interesting problems is that it forces them to think more about their answers and their own problem solving." Another teacher appreciated the fact that a discussion of responses to certain tasks fosters an environment in the classroom in which "students can begin to feel comfortable sharing what they think mathematically with … their classmates." At least one teacher was "surprised, at times, by some of the unique solution strategies that appear."

Activity 2: Editing Responses to Improve Their Quality

MATHEMATICS CLASSROOMS are changing from environments in which students provide answers without any justification to environments in which students are asked to solve mathematical tasks that require them to demonstrate their thinking and reasoning by providing explanations and justifications for their solutions. For students to become familiar with these kinds of tasks and to ensure that they do their best work, they need to be involved in a variety of activities that give them opportunities to examine and discuss the nature of the responses to the tasks. In activity 1, students explored multiple approaches to solving a problem. In activity 2, students also examine responses, but here, the discussion focuses on what elements constitute a complete and correct mathematical response. After talking about the requirements for high-level work, students examine lower-level responses and determine how they can be improved. Critiquing responses and writing suggestions for improving others' responses helps students become more reflective of their own problem solving and ensures that their own work and explanations are complete and correct.

Purpose of the activity

Students begin this activity by examining several responses that have been scored at the highest level. Once they have a shared understanding of the elements of a high-quality response, they are asked to examine a set of lower-level responses and determine whether each response is incorrect, incomplete, or both by identifying the problematic aspects of the response. Students then edit or rewrite the response to make sure that it is of high quality.

Description of the mathematics task

This task is the same task that was used in activity 1, the Blocks task. Any of the other tasks could be used for this purpose because they all result in differing levels of responses and because the lower-level responses can be improved. Teachers may want to select a task from the mathematics content that is currently being covered in their classroom. For example, teachers who are focusing on data analysis and probability might use the Left-Handed Survey, Jamal's Survey, Spinner, or Bar Average tasks.

The Blocks task (see **fig. 3.1**) involves number sense and basic concepts of number theory. Students must solve for an unknown number that satisfies several conditions set in a story context. Specifically, students must find the total number of blocks in a set given that one block remains when the whole set is partitioned into groups of two, three, or four. A condition of the problem is that the same total number of blocks is partitioned each time.

Implementation in the classroom

If students are not already familiar with the task, they should spend time solving it on their own. After working on the task for five or ten minutes, students can be asked to share their solutions with the rest of the class.

Discussing the elements of a complete and correct response

Before students are able to improve low-level responses, they need to have a good understanding of the requirements for a high-level response. For the Blocks task, the criteria for a score of 4 state that the work "shows a correct and complete understanding of common multiples and of the conditions of the task. The two conditions in the task are: (1) the same total number of blocks is partitioned each time into groups of two, three, and four; and (2) the correct numerical answer has a remainder of 1 when divided by 2, 3, and 4." Responses A through D were all scored at the highest level. For example, response A (**fig. 3.2**) completely and correctly applies the concept of common multiples to answer the question and satisfies the two conditions of the task. See activity 1 for a more detailed discussion of the multiple approaches and answers to this task.

All the tasks included in this book were scored using a 0-to-4-point rubric. For the highest-level score, the criteria in the QCAI general rubric indicate that the solution "identifies all the important elements of the problem and shows an understanding of the relationships between them"; however, rubrics vary in the criteria assigned to the highest level. For example, in some rubrics, the highest score possible is assigned only to responses that include generalizations. Talking with students about these differences in rubrics may be helpful. One way to start this kind of discussion is to look at the question asked in the Blocks task. What score would the students assign to a response that included more than one answer, such as thirteen and twenty-five, or to a response, such as response D (**fig. 3.5**), that explained that many possibilities exist for the number of blocks? Would students want to differentiate this response from the "typical 4" responses and, instead, assign it a score of 4*? Teachers might also ask students how they would score the Blocks task if the wording of the question was changed to ask, "How many different numbers of blocks could Yolanda have had?" Would the original criteria for a score of 4 need to be changed? Would solutions similar to those in responses A through C still be scored as 4s?

Editing responses

When students are comfortable recognizing the qualities of a level-4 response to the Blocks task, they can be divided into groups and asked to look at sample responses scored at the 3, 2, 1, and 0 levels, in this example, responses E through P. If specific score levels are to be discussed with students, teachers may want to show the scores with the responses; however, a discussion of the criteria at each score level is not necessary in this activity. Teachers could remove the assigned scores and simply refer to the packet as a set of responses that contain errors or may be incomplete. Note that the complete task packets contain rationales for the score given to each response. The rationale identifies why the response received the particular score that it was assigned.

Some teachers use a chart similar to the one in **figure 3.7** as a means for students to record group discussions. An advantage of using a chart or other recording mechanism is that it often encourages more focused interactions among the group members. For each response, students are first asked to indicate if the response is incorrect, incomplete, or both. Errors, misunderstandings, confusion, or lack of clear explanations are identified. Then, the response is "edited" to improve its quality. Students describe the changes, corrections, or additions they would make to the response and the reasons that the response was considered to be at a low level.

Discussing the edits and rationales

After the groups edit all responses, a worthwhile activity is to bring the whole class together to (*a*) allow groups to compare their edits and talk about the similarities and differences in how they thought the responses could be improved, and (*b*) bring the discussion to a more general level by identifying some overarching reasons for assigning low scores, such as "not knowing how to begin the problem," "using incorrect strategies," "making mathematical computation errors," "not meeting all the conditions of a problem," or "not being clear enough in the mathematical explanation."

The following paragraphs describe a discussion that students in Mr. Williams's class had when they were examining two low-level responses to this released task. Response H (**fig. 3.8**) shows a series of sketches to represent groups of blocks. One group of Mr. Williams's students determined that all three groupings have the same total number of blocks, nineteen, but the use of this total precludes satisfying the other condition of the task, which is to have one block remaining. The number 19 worked only for groups of two and three but not for groups of four. The student editors suggested using another total number of

Response	Incorrect, Incomplete, or Both	Changes, Corrections, or Additions to the Response	Reasons for Corrections or Additions

Fig. 3.7. Sample of a chart for recording group discussions

H   2

Yolanda was telling her brother Damian about what she did in math class.

Yolanda said, "Damian, I used blocks in my math class today. When I grouped the blocks in groups of 2, I had 1 block left over. When I grouped the blocks in groups of 3, I had 1 block left over. And when I grouped the blocks in groups of 4, I still had 1 block left over."

Damian asked, "How many blocks did you have?"

What was Yolanda's answer to her brother's question?

Show your work.

Groups of 2 →

Groups of 3 →

Answer: ___19___

Fig. 3.8 Activity 2—response H

blocks, such as thirteen or twenty-five, to ensure that all groupings would have one block left over.

Another example, response L (**fig. 3.9**), shows single groups of two, three, and four, each with one block left over. Students in one group reported to the rest of the class that the answer of 12 was merely the total number of blocks drawn in all three groupings. The only understanding shown in this response is that one block must be left over. The same number of blocks was not used in each grouping, and the answer was obtained using an incorrect method for this problem. The students in Mr. Williams's class talked about how they would explain to the student what was wrong with his or her solution.

Activity 3: Using Established Scoring Criteria to Evaluate the Quality of Responses

JUDGMENTS ABOUT the quality or adequacy of a solution or an explanation produced for a mathematics assessment are usually made by a teacher or an outside evaluator. Students themselves, however, can be evaluators of their own mathematical performances. In fact, according to *Assessment Standards for School Mathematics* (NCTM 1995), when students are adept at judging the quality of their own work or the work of others, they are more likely to demonstrate a complete understanding of the mathematics and to communicate their knowledge effectively. Thus, students can benefit from experiences in which they are asked to reflect on the quality of responses to mathematics tasks that require explanations or justifications. Activity 3 involves students in using established scoring criteria to score a set of responses to a task. After using this activity in their classrooms, teachers often found value in conducting similar activities using scoring rubrics from state, county, district, or school assessments to familiarize students with the expectations set forth in those rubrics.

Purpose of the activity

The activity described here familiarizes students with established scoring criteria for one of the tasks in this booklet. Students are shown the score-level headings and examples of responses at each of the five scoring levels, 0 to 4. They are then asked to apply the criteria to evaluate a set of responses to the task. Teachers have found that these kinds of activities encourage students to become more thoughtful when solving tasks, to give explanations and justifications for their solutions, and to recognize the importance of responding as com-

pletely and clearly as possible to enable the person reading the response to understand and evaluate the thinking behind it.

Description of the mathematics task

The task used in this activity is Miguel's Homework (see **fig. 3.10**). In conducting this activity, teachers must make sure that the majority of their students have at least some understanding of the mathematical content assessed by the task. If students have difficulty solving the task, their lack of understanding of the mathematics could impair their ability to examine and compare differing levels and qualities of responses. Miguel's Homework requires students to examine and recognize the underlying structure of a visual pattern. In part A, students are asked to continue the pattern by drawing the next figure for Miguel, and in part B, students are asked to describe to Miguel how they knew which figure came next in the pattern.

Implementation in the classroom

To familiarize themselves with the task itself and the content that is being assessed, students should solve the task on their own, then share their solutions with the class.

Discussing the established scoring criteria

The class may begin by discussing what is expected for a complete and correct response to the task. As the score-level headings for a 4 indicate, two aspects of this pattern must be detected and described completely and correctly: (1) the number of dots, and (2) the shape of the figure or the placement of the dots to be added. Using the scored responses and rationales will help clarify the scoring criteria at each level. For example, responses A, E, F, J, and L can be used to illustrate score levels of 4, 3, 2, 1, and 0, respectively.

Response A (**fig. 3.11**): This student drew the next figure in the pattern correctly. The explanation completely and correctly states the number of dots added to each figure and indicates that the dots are added to the side of each figure. For this reason, the response is assigned a score of 4.

Response E (**fig. 3.12**): The correct figure is provided in this response, but the explanation does not specifically indicate that one more dot was added to each row of the figure. Thus, the response received a score of 3.

Yolanda was telling her brother Damian about what she did in math class.

Yolanda said, "Damian, I used blocks in my math class today. When I grouped the blocks in groups of 2, I had 1 block left over. When I grouped the blocks in groups of 3, I had 1 block left over. And when I grouped the blocks in groups of 4, I still had 1 block left over."

Damian asked, "How many blocks did you have?"

What was Yolanda's answer to her brother's question?

Show your work.

Answer: ___12___

Fig. 3.9 Activity 2—response L

Parke, Carol S., Suzanne Lane, Edward A. Silver, and Maria E. Magone. *Using Assessment to Improve Middle-Grades Mathematics Teaching and Learning: Suggested Activities Using QUASAR Tasks, Scoring Criteria, and Students' Work.* Reston, Va.: National Council of Teachers of Mathematics, 2003.

Miguel's Homework Task

For homework Miguel's teacher asked him to look at the pattern below and draw the figure that should come next.

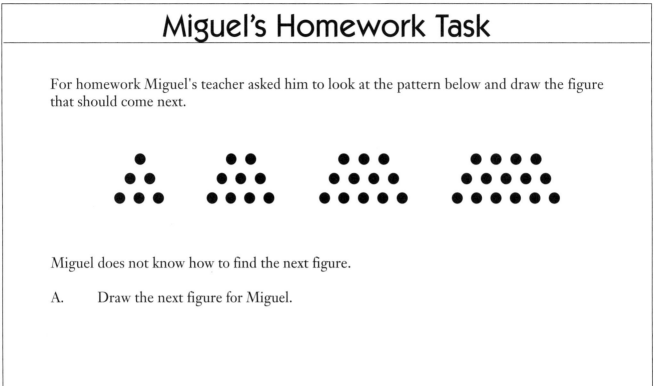

Miguel does not know how to find the next figure.

A.   Draw the next figure for Miguel.

B.   Write a description for Miguel telling him how you knew which figure comes next.

Fig. 3.10. Activity 3—Miguel's Homework task

For homework Miguel's teacher asked him to look at the pattern below and draw the figure that should come next.

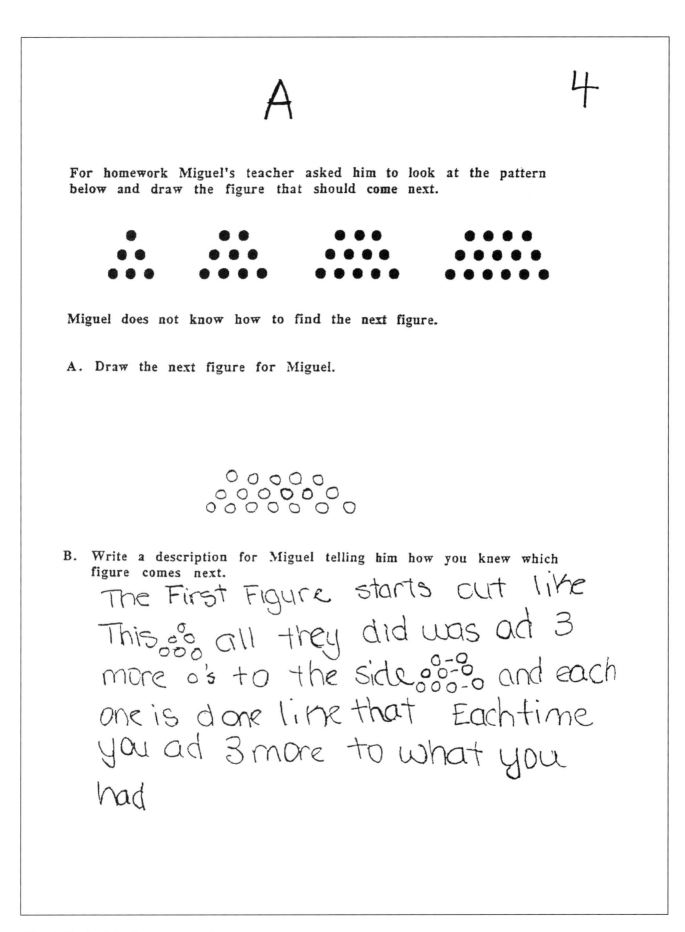

Miguel does not know how to find the next figure.

A. Draw the next figure for Miguel.

B. Write a description for Miguel telling him how you knew which figure comes next.

The First Figure starts out like This all they did was ad 3 more o's to the side and each one is done like that Each time you ad 3 more to what you had

Fig. 3.11. Activity 3—response A

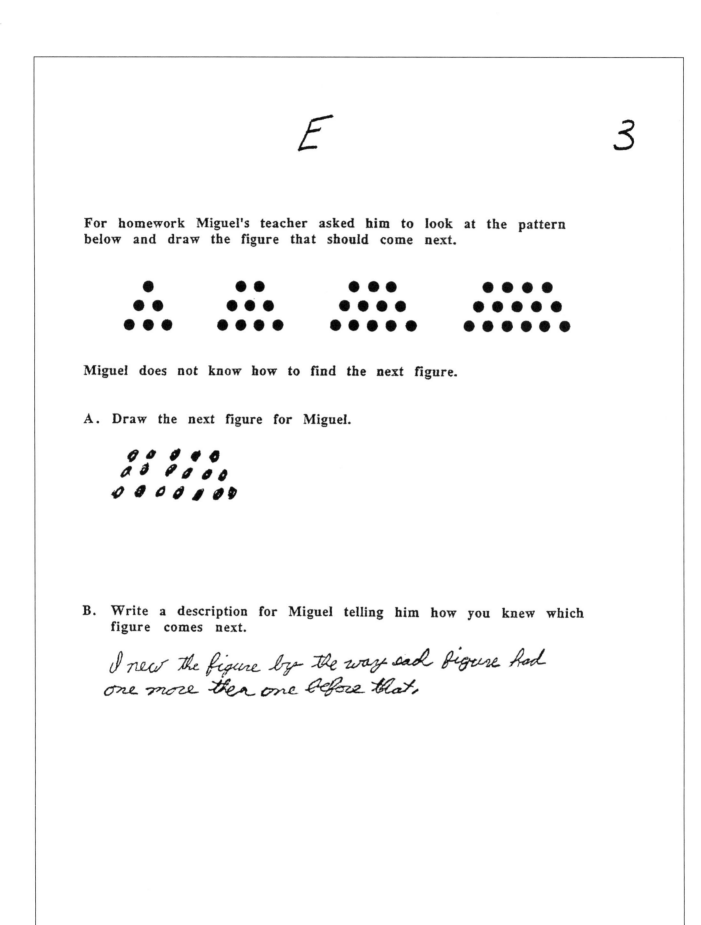

For homework Miguel's teacher asked him to look at the pattern below and draw the figure that should come next.

Miguel does not know how to find the next figure.

A. Draw the next figure for Miguel.

B. Write a description for Miguel telling him how you knew which figure comes next.

I new the figure by the way sad figure had one more then one before that.

Fig. 3.12. Activity 3—response E

Response F (**fig. 3.13**): This response can be used to illustrate a score of 2. The figure contains the correct number of dots but not the correct shape. In the description of the pattern, the response partially identifies the pattern by indicating the correct increase in the number of dots but is incomplete because it does not explain where the three dots are added.

Response J (**fig. 3.15**): Response J has no explanation, and the figure is not completely correct. It received a 1 instead of a 0, however, because the figure shows the correct number of dots in the first and third rows.

Response L (**fig. 3.16**): This response received the lowest score level, a 0. The response offers no description of the pattern, and the figure is simply a redrawing of the fourth figure, which is already provided in the task.

Scoring responses

Groups of students should be given sets of approximately eight to ten unscored responses. One option would be to use responses from the packet that the class has not discussed. The responses should be placed in random order, and the scores should be hidden. Responses B, C, D, G, H, I, K, and M in this book cover the range of score levels. Another option would be to use your own students' responses with all names removed.

Ask students to work in groups to (*a*) discuss each response, (*b*) apply the criteria and come to an agreement on the score for each response, and (*c*) write a brief rationale for the assigned score. Some teachers use a chart, such as the one shown in **figure 3.17**, for students to record their scores and rationales.

Discussing students' scores

When students finish scoring all responses, tally the groups' scores for each response for the whole class to see. If the responses in the packet are used, the actual scores that were assigned to the responses may also be shared with students at this point. A whole-class discussion of the scores could be conducted in many ways. A teacher might begin by discussing the responses in the order they were given to students. Alternatively, a teacher could look at the tallies of assigned scores and begin by focusing on responses that were assigned the same score by all groups, that is, responses on which students reached a high level of agreement. Groups of students can give their rationales for the score and talk about why they thought the response was easy to score or why it seemed to fit so clearly into one of the scoring categories. Then the discussion can turn to those responses on which students disagreed, and students can identify the characteristics of the response that might have led to differences in scores. The following paragraphs illustrate the nature of the interactions that can take place among students and teachers when discussing disagreements in scores for Miguel's Homework.

In Ms. Richardson's class, response I (**fig. 3.14**) was scored as a 2, a 1, and a 0 by different groups of students. This response shows the correct number of dots in the figure, but the shape of the figure is incorrect. Ms. Richardson asked the groups to give rationales for their scores. Tamara said that her group assigned the response a 0 because "The drawing is wrong; there should not be six dots in each row." This group of students realized that the shape was incorrect; however, they did not notice that the total number of dots in the figure was correct. Calvin's group assigned the response a score of 2, and the group members said that they compared it with response F (**fig. 3.13**), which Ms. Richardson had used as an example of a 2 response: "Both responses have a figure that is wrong because of the shape, but the number of dots is right." Billy's group disagreed with this comparison, pointing out, "Response F describes part of the pattern and labels the total number of dots on each figure, but response I only draws the figure. That's why we gave it a 1."

Ms. Richardson was pleased with the rich discussions that this activity fostered among her students. As the groups talked about their rationales, they referred to the score-level headings and previously discussed responses. This background information was valuable in helping students gain a better understanding of the scoring criteria. After each group shared its rationale, the class agreed that response I seemed to fit best in the score level of 1.

Response	Score	Rationale for Score

Fig. 3.17. Sample chart for recording scores and rationales

For homework Miguel's teacher asked him to look at the pattern below and draw the figure that should come next.

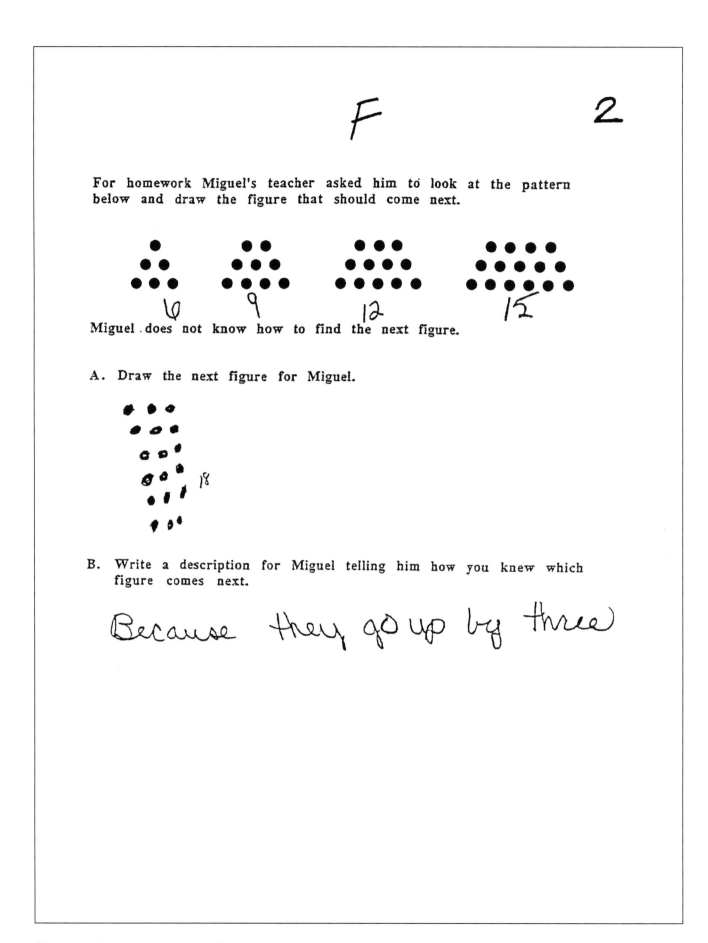

Miguel does not know how to find the next figure.

A. Draw the next figure for Miguel.

B. Write a description for Miguel telling him how you knew which figure comes next.

Because they go up by three

Fig. 3.13. Activity 3—response F

I 1

For homework Miguel's teacher asked him to look at the pattern below and draw the figure that should come next.

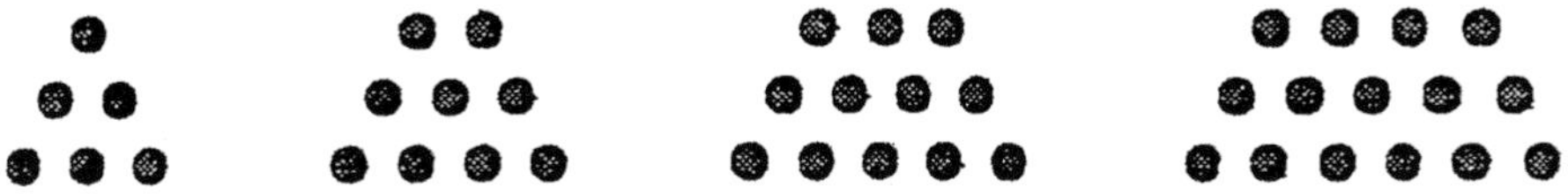

Miguel does not know how to find the next figure.

A. Draw the next figure for Miguel.

B. Write a description for Miguel telling him how you knew which figure comes next.

Fig. 3.14. Activity 3—response I

For homework Miguel's teacher asked him to look at the pattern below and draw the figure that should come next.

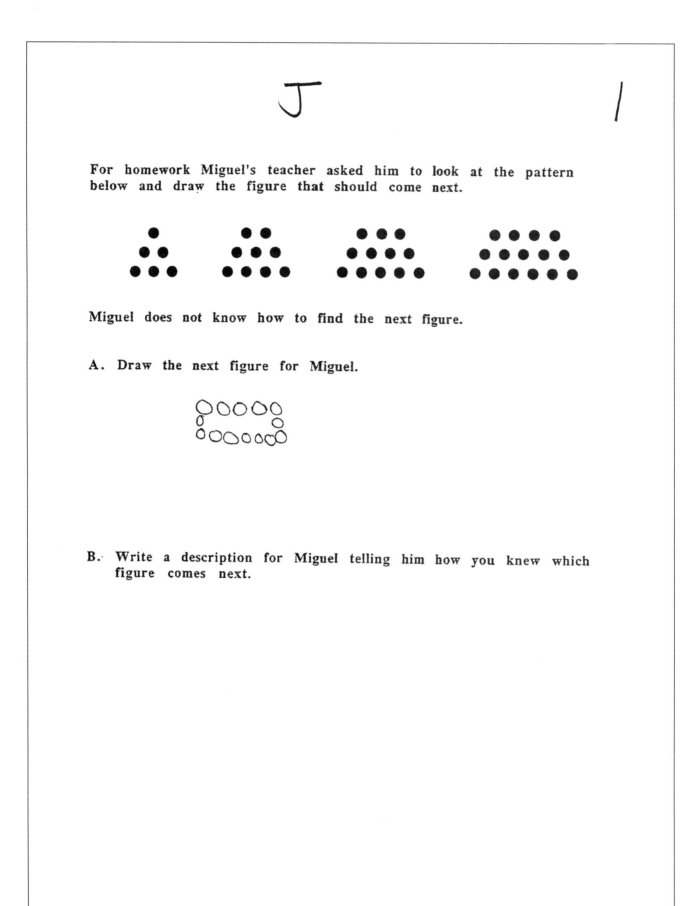

Miguel does not know how to find the next figure.

A. Draw the next figure for Miguel.

B. Write a description for Miguel telling him how you knew which figure comes next.

Fig. 3.15. Activity 3—response J

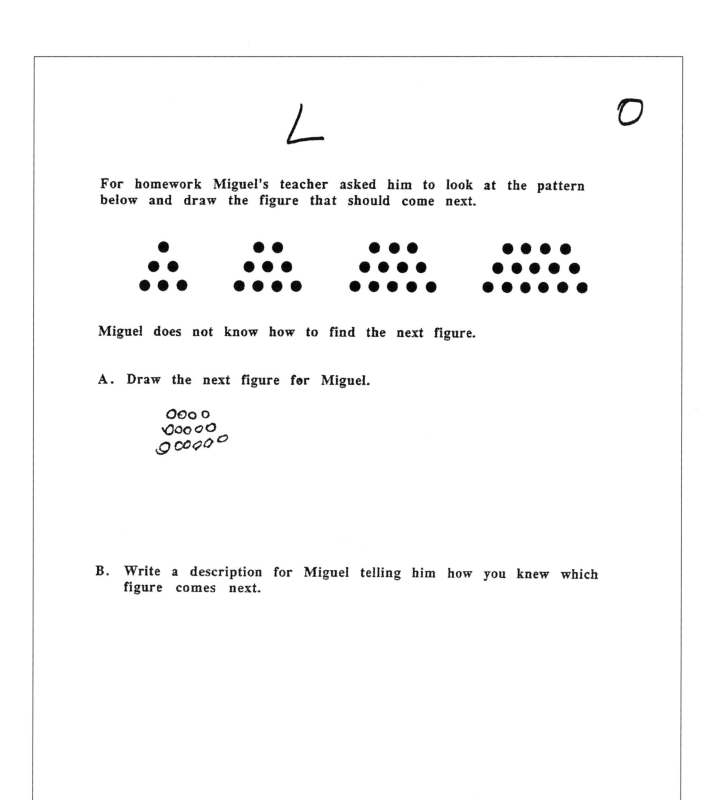

L

O

For homework Miguel's teacher asked him to look at the pattern below and draw the figure that should come next.

Miguel does not know how to find the next figure.

A. Draw the next figure for Miguel.

B. Write a description for Miguel telling him how you knew which figure comes next.

Fig. 3.16. Activity 3—response L

Activity 4: Developing Scoring Criteria to Evaluate the Quality of Responses

As STUDENTS evaluate their own work or the work of others, they are crystallizing their ideas about the features of a high-quality mathematical response to a complex problem. In activity 3, students used established scoring criteria to judge the quality of responses. In activity 4, students again evaluate the level of understanding demonstrated in responses, but they now develop their own set of criteria. This activity gives students opportunities to reflect critically on the desirable qualities of a mathematical response by examining several responses to a task and differentiating among them using their own judgments. Watching students and listening to them interact to develop criteria, teachers gain valuable insights into their students' ideas about what constitutes good mathematical problem solving and reasoning.

Purpose of the activity

In this activity, students judge the quality of responses to a mathematical task. They are given a set of unscored responses to a task and are asked to differentiate among them using their own judgments. Students sort responses into three categories, high, midlevel, and low, according to the level of mathematical understanding demonstrated. Throughout this process, students discuss the reasons that a response is placed in a particular level, then reach consensus about the criteria for each of the three levels.

Description of the mathematics task

This activity uses the Bar Average task (see **fig. 3.18**). Again, teachers must make sure that the majority of students understand the mathematical content of any task chosen for this activity. If students have difficulty solving the task, their lack of understanding of the mathematics could interfere with their ability to examine and compare the quality of responses. The Bar Average task assesses students' understanding of the concept of average. Given a bar graph showing a student's first three science project scores and the average of all four scores, students are required to find the missing fourth score. Students can use different solution methods to determine the answer, including arithmetic calculations, a guess-and-check strategy, graphic representations, and algebraic approaches.

Implementation in the classroom

To familiarize themselves with the task and the content that is being assessed, students should solve the task individually, then share their solutions with the class.

Developing criteria and scoring responses

To begin the activity, distribute a packet of materials to each group. This packet could include a set of approximately eight to ten randomly ordered sample responses with the scores hidden, a rating sheet for the students to record their scores, and a sheet for recording the criteria developed by the group. A rating sheet might simply have two columns: one to identify the response and another for the score assigned by the group. The criteria sheet could have three sections: one each to describe the criteria for high-level, mid-level, and low-level responses.

In their groups, student should (*a*) read and examine all responses carefully before assigning any scores, (*b*) sort responses into three levels—low, midlevel, and high—on the basis of the quality of the responses, (*c*) record scores on a rating sheet, and (*d*) write descriptions of the criteria developed for each level. Students could sort the responses into more than three levels; however, for their first experiences in developing criteria, three levels might be easier to differentiate.

Discussing students' criteria and scores

After students complete the scoring activity, tally the groups' scores at each level for each response. Recording the tallies for all the groups will allow the class to examine the agreements and disagreements in the evaluations of responses across the groups. A whole-class discussion of the scoring-activity results could proceed in a number of ways. In general, a teacher might want to begin with a comparison of how each response was scored, then ask students to share the criteria they generated for each level. The following paragraphs highlight interactions in Mr. Collins's classroom to illustrate the rich and lively mathematical discussions that can take place among students and teachers when they are developing and sharing their criteria for levels of responses.

Mr. Collins began the class discussion with a response that everyone agreed should be placed in the low-level category, response K (**fig. 3.22**). When Mr. Collins asked students to explain their reasoning for assigning this response a low-level score, Bernadette said, "The answer is wrong, and the explanation just didn't make any sense to me. It says, 'Count the

Parke, Carol S., Suzanne Lane, Edward A. Silver, and Maria E. Magone. *Using Assessment to Improve Middle-Grades Mathematics Teaching and Learning: Suggested Activities Using QUASAR Tasks, Scoring Criteria, and Students' Work.* Reston, Va.: National Council of Teachers of Mathematics, 2003.

Bar Average Task

Anita has four 20-point projects for science class. Anita's scores on the first three projects are shown below.

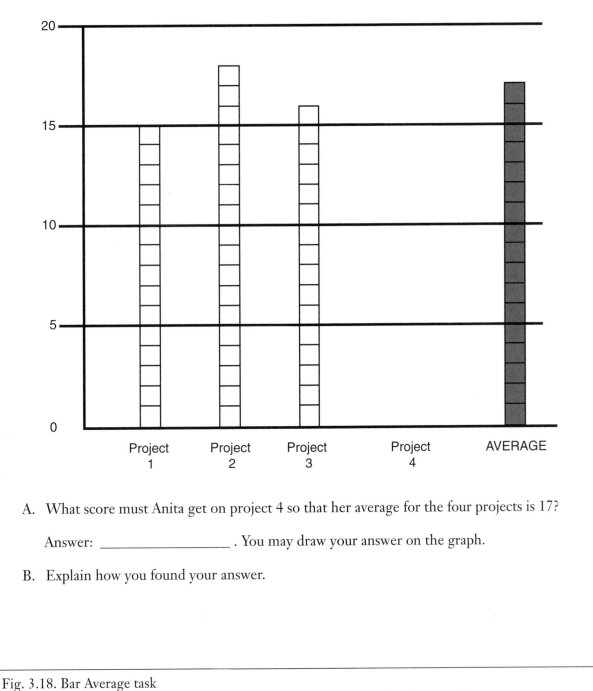

A. What score must Anita get on project 4 so that her average for the four projects is 17?

   Answer: _____ . You may draw your answer on the graph.

B. Explain how you found your answer.

Fig. 3.18. Bar Average task

blocks,' but that doesn't really explain anything." Next, Mr. Collins asked his students to share the criteria they generated for classifying low-level responses. Bernadette said, "We wrote that a low is … 'wrong answer; explanation doesn't make sense.'" Another student, Colleen, said that her group decided that a response, even one with a correct answer, should be categorized as low level if no explanation or sketch was provided. Other students contributed additional ideas. At the end of this discussion, a complete list of students' criteria for classifying a response as low level was shown on the overhead projector.

Next Mr. Collins turned students' attention to criteria for classifying responses in the highest category and asked one group to explain why response A (**fig. 3.19**) was classified as a high-level response. One student, Ken, observed, "The answer was right, and the sketch showed how the answer was found. See, one block from project 2 can be given to project 3. Now both of those projects have seventeen. But project 1 still has two less than seventeen, so if you add two more blocks above seventeen for project 4, then those two blocks can be put onto project 1. Then all the projects can average out to seventeen." Ken said that he and his group decided that a high–level response needed to "have a correct answer, correct explanation, and correct sketch." Mr. Collins asked if everyone else agreed that a high-level response needed to meet all these criteria. Jackie raised her hand hesitantly and said, "We kind of have the same thing because we think the answer needs to be correct, but we thought it didn't matter if there was a sketch as long as the explanation was complete." As an example of a response that her group judged at a high level, Jackie referred to response C (**fig. 3.20**).

The students in Jackie's group thought that this response showed a good way to solve the problem even though no sketch was used. Because some students did not understand the method of solution in response C, Mr. Collins asked Jackie to explain it. She said, "First, they found all the points Anita got so far; that was 49. Then they found out how many points it would be if all four projects were given the average of 17. That was 68. Then they subtracted the 49 points Anita already had from the 68 total points, so Anita needed 19 points." After listening to the explanation of this approach to the task, other students began to realize that a sketch might not be necessary if a solution process such as this one was used. As other groups shared their criteria for high-level responses, Mr. Collins again summarized them on the overhead projector.

Finally, Mr. Collins discussed the midlevel responses with his class. Students wanted to talk about response H (**fig. 3.21**). Raoul said, "The answer is wrong. This person did almost what the person in response C did, but they didn't do everything right. They got the number 49, but you can't tell what some of the other steps were." Other students agreed and described the criteria for a midlevel response as "wrong answer; some parts of the work are correct, but it's hard to follow and incomplete."

Teachers who have used this activity often comment that they are glad to see their students' intense involvement and interaction during the group work and whole-class discussion. Teachers have also commented that they learn a lot about their students from this activity because it brings to the forefront students' impressions of the elements of good mathematical thinking.

Activity 5: Assessing Students' Existing Knowledge

STUDENTS' EXPLANATIONS, work, and answers to performance tasks are valuable sources of information for determining what they know about a particular mathematical content area. Before introducing a new topic in the classroom, teachers can assess students' existing knowledge by examining responses to a small set of tasks designed to ascertain that knowledge. Instruction can then be tailored to the specific needs of students. In this description of activity 5, a task is used that assesses students' existing knowledge of area and perimeter, as demonstrated by their solutions to the task.

Purpose of the activity

This activity illustrates how responses to tasks can inform teachers about their students' knowledge of a mathematical content area before instruction. On completion of the QUASAR project, all the QCAI tasks were made available to teachers, giving them an opportunity to use several tasks covering a particular mathematical topic before beginning an instructional unit in that content area. When the purpose of examining students' responses is to guide instruction, teachers do not typically assign formal scores to the responses; instead, they explore the responses and study the various ways that students approach the tasks and the types of understanding and misunderstanding, areas of competence and weakness, and levels of proficiency and confusion that students display in their work.

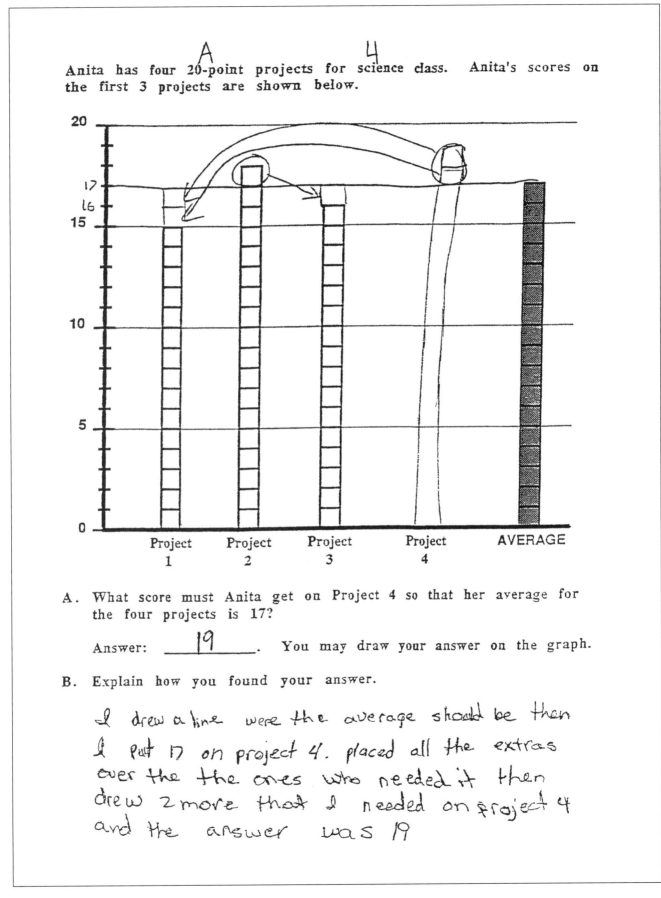

Anita has four 20-point projects for science class. Anita's scores on the first 3 projects are shown below.

A. What score must Anita get on Project 4 so that her average for the four projects is 17?

Answer: ___19___. You may draw your answer on the graph.

B. Explain how you found your answer.

I drew a line were the average shoud be then I put 17 on project 4. placed all the extras over the the ones who needed it then drew 2 more that I needed on project 4 and the answer was 19

Fig. 3.19. Activity 4—response A

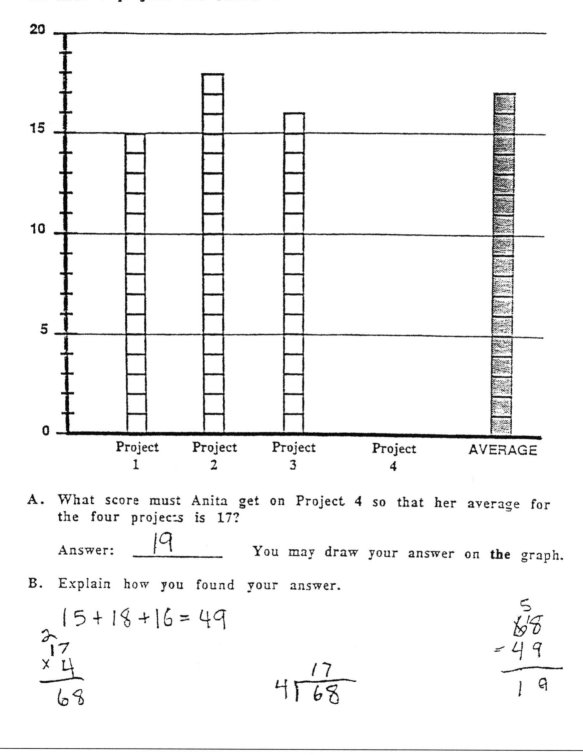

C   4

Anita has four 20-point projects for science class.   Anita's scores on the first 3 projects are shown below.

A. What score must Anita get on Project 4 so that her average for the four projects is 17?

Answer: _____19_____   You may draw your answer on **the** graph.

B. Explain how you found your answer.

$15 + 18 + 16 = 49$

$\begin{array}{r} 2 \\ 17 \\ \times\ 4 \\ \hline 68 \end{array}$

$4\overline{)68}^{\,17}$

$\begin{array}{r} 5 \\ \cancel{6}\cancel{8} \\ -4\ 9 \\ \hline 1\ 9 \end{array}$

Fig. 3.20. Activity 4—response C

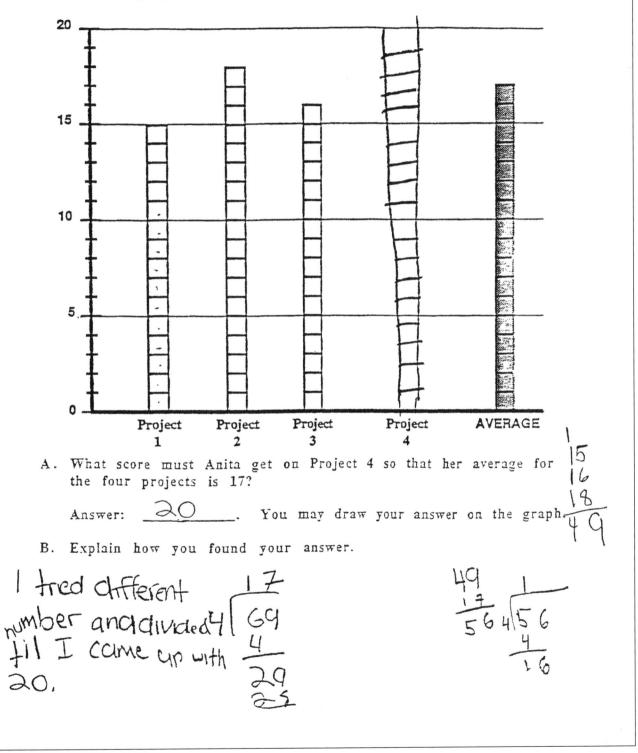

Anita has four 20-point projects for science class. Anita's scores on the first 3 projects are shown below.

A. What score must Anita get on Project 4 so that her average for the four projects is 17?

Answer: ___20___. You may draw your answer on the graph.

B. Explain how you found your answer.

I tried different number and divided til I came up with 20.

$$
\begin{array}{r}
17 \\
4\overline{\smash{)}69} \\
4 \\
\hline
29 \\
25
\end{array}
$$

$$
\begin{array}{r}
49 \\
17 \\
\hline
56
\end{array}
\qquad
\begin{array}{r}
1 \\
4\overline{\smash{)}56} \\
4 \\
\hline
16
\end{array}
$$

$$
\begin{array}{r}
15 \\
16 \\
18 \\
\hline
49
\end{array}
$$

Fig. 3.21. Activity 4—response H

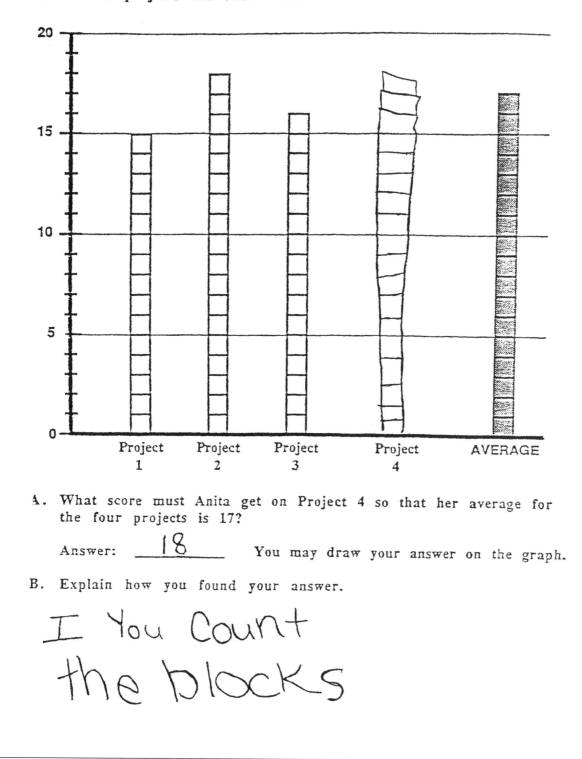

K   O

Anita has four 20-point projects for science class.   Anita's scores on
the first 3 projects are shown below.

A. What score must Anita get on Project 4 so that her average for
   the four projects is 17?

   Answer: ___18___   You may draw your answer on the graph.

B. Explain how you found your answer.

I You Count
the blocks

Fig. 3.22. Activity 4—response K

Description of the mathematics task

The School Board task (see **fig. 3.23**), which measures knowledge of area and perimeter, is used as an example in this activity. Four additional tasks in this book, Art Project, Art Class, Mr. Jackson's Fence, and Double the Carpet, can be grouped as a set with the School Board task to assess several important aspects of students' knowledge of area and perimeter. Sets of tasks covering other content areas could also be used. For example, this book includes four tasks dealing with numbers and operations, four tasks in data analysis and probability, and three tasks in algebra, functions, and patterns.

The School Board task assesses whether students can find the area and perimeter of two rectangular shapes. Interestingly, the task does not use the terms *area* and *perimeter*. Rather, students are asked to find the piece of land that covers "as much land as possible," or has the largest area, and the piece of land that requires "less fencing," or has the smallest perimeter. That is, without using technical vocabulary, students are being asked to identify the shape with the largest area and the shape with the smallest perimeter.

Information obtained from students' responses

The following paragraphs describe the insight gained by one teacher, Ms. Malloy, from using the School Board task. After examining responses from her class, Ms. Malloy noted several areas that would require attention in instruction. A few of Ms. Malloy's students seemed to have a good grasp of how to find the area and perimeter of a rectangular shape. They knew when to apply the area and perimeter formulas and how to calculate them correctly. The students were sometimes careless, however, in interpreting what the question was asking. For example, the work shown in response F (**fig. 3.24**) is correct in both parts, but in part B, the student chose the land with the larger perimeter rather than the land that required the least amount of fencing.

In another response, similar to the one in response G (**fig. 3.25**), part A had an incorrect answer with no work. Part B showed the correct answer, along with some relevant work. Written calculations showed that the student had added the lengths and widths of the two figures, but no other work or explanation was shown, and Ms. Malloy wondered whether the student truly understood the concept of perimeter. Because the task did not ask for the perimeters, the question could simply be answered by adding the length and

width for each figure. Ms. Malloy decided to question students when she sees this type of explanation to ensure that they truly understand how to find the perimeter. Students might think that if the area is found by multiplying the length and width, then the perimeter is found by adding the length and width; they may not realize that they find only half of the perimeter with this method.

Other students seemed to know the formulas for finding area or perimeter or both but confused them or applied them incorrectly. Some students, such as the one who submitted response H (**fig. 3.26**), used the perimeter formula to find the answer to both parts. Other students used the perimeter formula to answer part A and the area formula to answer part B. They seemed to memorize formulas without understanding what the calculations represent conceptually or how to apply them.

Finally, one student used no formulas or calculations but wrote an explanation to justify the answer. In response L (**fig. 3.27**), the explanation in part A seems to show the beginnings of a conceptual understanding of increasing areas. As the student notes, if both pieces of land have the same perimeter, then the closer the shape gets to a square, the larger the area becomes. Ms. Malloy thought that she could use this type of response later in her instruction and have students discuss its merit. Then students could work on an activity to examine relationships between the area and perimeter of rectangles.

Activity 6: Monitoring Students' Learning during Instruction

PERIODICALLY ASSESSING students using performance tasks throughout instruction can help teachers diagnose students' misunderstandings and make instructional decisions. Both teachers and students benefit from the information learned in monitoring students' progress. Specific feedback can be given to students about their levels of understanding. This feedback involves more than simply telling students whether they have gotten the correct answer. Instead, students can find out exactly why their responses were incomplete or incorrect and where they might have gone astray in the thinking process. This discussion of activity 6 also outlines a task involving area and perimeter to illustrate how teachers can use such tasks to tailor instruction and give students information about their levels of understanding of particular topics.

Parke, Carol S., Suzanne Lane, Edward A. Silver, and Maria E. Magone. *Using Assessment to Improve Middle-Grades Mathematics Teaching and Learning: Suggested Activities Using QUASAR Tasks, Scoring Criteria, and Students' Work*. Reston, Va.: National Council of Teachers of Mathematics, 2003.

School Board Task

The school board wants to buy a piece of land. Mrs. Gomez and Mr. Langer are each selling a rectangular piece of land next to the school.

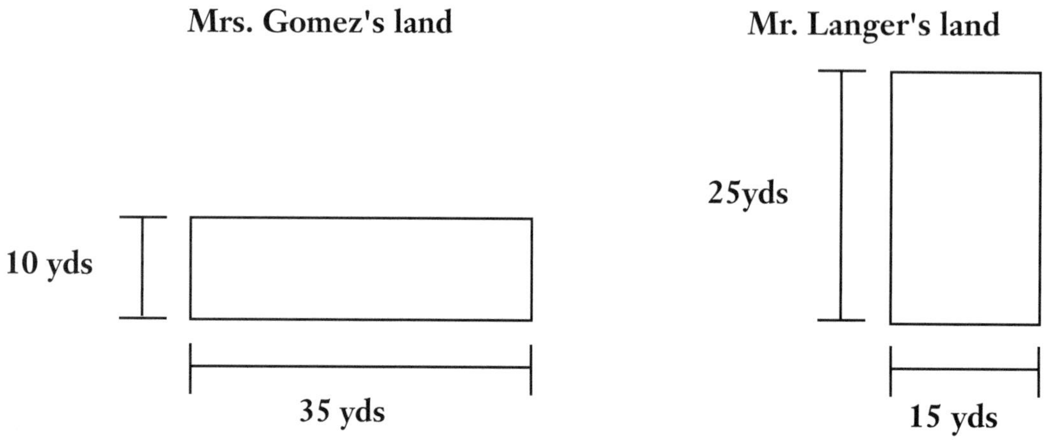

Mrs. Gomez's land

Mr. Langer's land

10 yds

35 yds

25yds

15 yds

A. If the school board wants <u>as much land</u> as possible, should the school board buy from Mrs. Gomez or Mr. Langer?

Show how you found your answer.

Answer: _____

B. If the school board wants to surround either Mrs. Gomez's land or Mr. Langer's land with a fence, which piece of land would require the <u>lesser amount</u> of fencing?

Show how you found your answer.

Answer: _____

Fig. 3.23. School Board Task

F 3

The school board wants to buy a piece of land. Mrs. Gomez and Mr. Langer are each selling a rectangular piece of land next to the school.

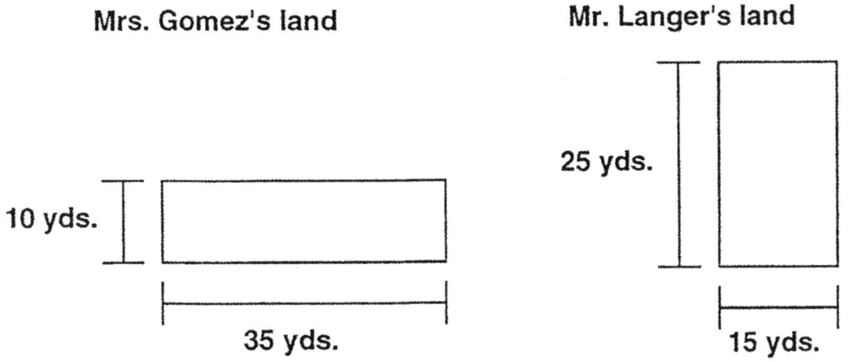

Mrs. Gomez's land

Mr. Langer's land

25 yds.

10 yds.

35 yds.

15 yds.

A. If the school board wants <u>as much land</u> as possible, should the school board buy from Mrs. Gomez or Mr. Langer?

Show how you found your answer.

$$
\begin{array}{r}
35 \\
\times\ 10 \\
\hline
0\ 0 \\
3\ 5 \\
\hline
350
\end{array}
$$

$$
\begin{array}{r}
25 \\
\times\ 15 \\
\hline
125 \\
+25 \\
\hline
375
\end{array}
$$

Answer: <u>Mr. Langers</u>

B. If the school board wants to surround either Mrs. Gomez's land or Mr. Langer's land with a fence, which piece of land would require the <u>least amount</u> of fencing?

Show how you found your answer.

Mrs. Gomez
$$
\begin{array}{r}
35 \\
35 \\
+10 \\
10 \\
\hline
90
\end{array}
$$

Mr Langer
$$
\begin{array}{r}
2 \\
25 \\
25 \\
15 \\
+15 \\
\hline
80
\end{array}
$$

Answer: <u>Mrs, Gomezs</u>

Fig. 3.24. Activity 5— response F

G 2

The school board wants to buy a piece of land. Mrs. Gomez and Mr. Langer are each selling a rectangular piece of land next to the school.

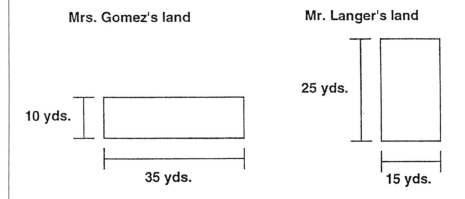

Mrs. Gomez's land

Mr. Langer's land

10 yds.

35 yds.

25 yds.

15 yds.

A. If the school board wants <u>as much land</u> as possible, should the school board buy from Mrs. Gomez or Mr. Langer?

Show how you found your answer.

Answer: <u>Mrs. Gomez's Land</u>

B. If the school board wants to surround either Mrs. Gomez's land or Mr. Langer's land with a fence, which piece of land would require the <u>least amount</u> of fencing?

Show how you found your answer.

Mrs. Gomez

$$\begin{array}{r} \overset{1}{1}0 \\ +\ 35 \\ \hline 45 \end{array}$$

Mr. Langers

$$\begin{array}{r} 25 \\ +15 \\ \hline 40 \end{array}$$

Answer: <u>Mr. Langers</u>

Fig. 3.25 Activity 5— response G

H 2

The school board wants to buy a piece of land. Mrs. Gomez and Mr. Langer are each selling a rectangular piece of land next to the school.

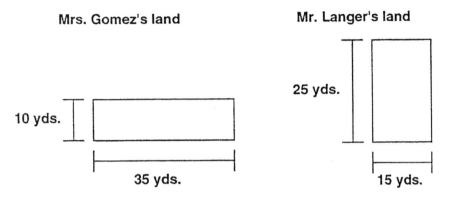

Mrs. Gomez's land

10 yds.

35 yds.

Mr. Langer's land

25 yds.

15 yds.

A. If the school board wants <u>as much land</u> as possible, should the school board buy from Mrs. Gomez or Mr. Langer?

Show how you found your answer.

$$
\begin{array}{cc} 10 & 2.5 \\ 35 & 15 \\ \hline 45 & 40 \end{array}
\qquad
\begin{array}{cc} 10 & 25 \\ 10 & 25 \\ 35 & 15 \\ 35 & 15 \\ \hline 90 & 80 \end{array}
$$

Answer: The school board should by Mrs. Gomez land.

B. If the school board wants to surround either Mrs. Gomez's land or Mr. Langer's land with a fence, which piece of land would require the <u>least amount</u> of fencing?

Show how you found your answer.

$$
\begin{array}{cc} 10 & 25 \\ 10 & 25 \\ 35 & 15 \\ 35 & 15 \\ \hline 90 & 80 \end{array}
$$

Answer: Mr. Langer land will take less fence.

Fig. 3.26. Activity 5— response H

L I

The school board wants to buy a piece of land. Mrs. Gomez and Mr. Langer are each selling a rectangular piece of land next to the school.

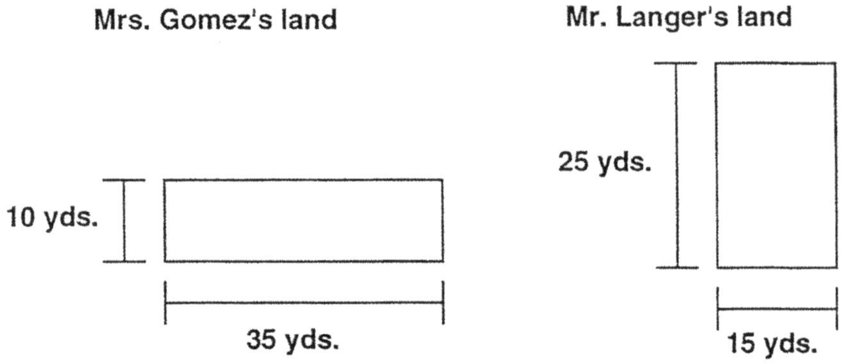

Mrs. Gomez's land

Mr. Langer's land

10 yds.

35 yds.

25 yds.

15 yds.

A. If the school board wants <u>as much land</u> as possible, should the school board buy from Mrs. Gomez or Mr. Langer?

Show how you found your answer. because Mr. Langer Land is kind of squarish and Mr. Gomez Land is reqtangular

Answer: _Mr. Langer's Land_

B. If the school board wants to surround either Mrs. Gomez's land or Mr. Langer's land with a fence, which piece of land would require the <u>least amount</u> of fencing?

Show how you found your answer. because his Land is thicker and you don't have to go to wide strechs of land

Answer: _Mr Gomez's Land_

Fig. 3.27. Activity 5— response L

Purpose of the activity

In this activity, a set of tasks that focuses on measurement was administered to all students near the end of an instructional unit. Teachers were given all the QCAI tasks on completion of the project and could use several tasks to assess the same mathematical content area. The following paragraphs explain how one teacher, Mr. Greystone, used the tasks to get a sense of his students' knowledge of perimeter and area during an instructional unit on these topics. Because he wanted to know the levels of students' understanding, Mr. Greystone scored the responses using the score-level headings in the task packets. After reviewing and scoring all responses, he shared the results with his students.

Description of the mathematics task

The Double the Carpet task (see **fig. 3.28**) is used to illustrate what Mr. Greystone learned about his students' knowledge of area and perimeter. Four additional tasks that assess important aspects of area and perimeter, Art Project, Art Class, Mr. Jackson's Fence, and School Board, are provided in this book. Sets of tasks covering other content areas could also be used for this purpose. For instance, this book includes four tasks dealing with numbers and operations, four tasks in data analysis and probability, and three tasks in algebra, functions, and patterns.

Double the Carpet assesses students' knowledge of the relationship between the perimeter and area of a figure. Students are given the dimensions of two rectangular rooms, with the perimeter of the game room twice as large as that of the living room, and are asked to determine whether the area of the game room is also twice as large. Students may use diagrams, written explanations, arithmetic calculations, or a combination of the three to support their answers.

Information obtained from students' responses

The following paragraphs describe the knowledge that Mr. Greystone and his students gained about the levels of students' understanding of area and perimeter. One student, whose work was similar to that shown in response A (**fig. 3.29**), had a complete and correct understanding of the relationship between perimeter and area. The sketch is complete, and the explanation goes beyond expectations by stating, "It [the game room] would be double the area if you only doubled one of the dimensions as you can see in my sketch." Mr. Greystone showed this response to the class because it was clearly a high-level response.

Another student, whose work is shown in response F (**fig. 3.31**), appeared to have a solid grasp of the concept of area as it was applied in this problem context. The one difficulty with this response, however, is that the area of the game room is actually four times the size of the living room, not three times the size. This student's response of "3 times the size" may be an erroneous representation of a part-part comparison (area of living room covers one part of game room, leaving three parts uncovered) rather than a part-whole comparison (area of living room is one of four equal parts of the game room).

In response E (**fig. 3.30**), both the answer and the calculations for the area are correct. Mr. Greystone might assume that a student who submitted this response has a complete understanding of area; however, the response offers no explicit or implicit comparison of the areas to justify the answer of "no." Mr. Greystone cannot be certain that the student understands why "no" is the correct answer. This student should work on explaining the thinking that led to the answer.

Finally, response M (**fig. 3.32**) displays a major misconception about the relationship between area and perimeter. The student thinks that if the perimeter is doubled, then the area must be doubled as well. Mr. Greystone might review instruction on area to address the deficiencies and misconceptions in these responses.

Parke, Carol S., Suzanne Lane, Edward A. Silver, and Maria E. Magone. *Using Assessment to Improve Middle-Grades Mathematics Teaching and Learning: Suggested Activities Using QUASAR Tasks, Scoring Criteria, and Students' Work.* Reston, Va.: National Council of Teachers of Mathematics, 2003.

Double the Carpet Task

Mr. Goldstein wants to buy carpet to cover the floors completely in his living room and game room. His living room is 10 feet by 15 feet. His game room is 20 feet by 30 feet.

Mr. Goldstein thinks that the area to be carpeted in his game room is <u>double</u> the area to be carpeted in his living room.

Is Mr. Goldstein correct?

Explain your answer, and show your work. You may use diagrams as part of your explanation and work.

Fig. 3.28. Double the Carpet task

4

A

Mr. Goldstein wants to buy carpet to cover the floors completely in his living room and game room. His living room is 10 feet by 15 feet. His game room is 20 feet by 30 feet.

Mr. Goldstein thinks that the area to be carpeted in his game room is <u>double</u> the area to be carpeted in his living room.

Is Mr. Goldstein correct?

Explain your answer and show your work. You may use diagrams as part of your explanation and work.

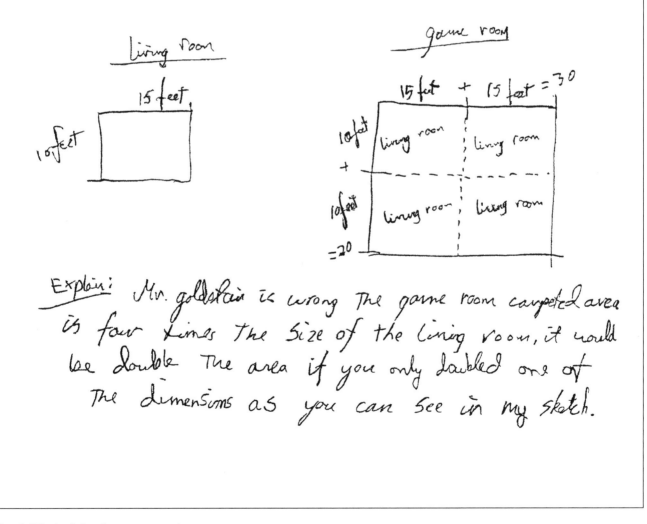

Explain: Mr. goldstein is wrong The game room carpeted area is four times The size of the living room, it would be double The area if you only doubled one of The dimensions as you can see in my sketch.

Fig. 3.29. Activity 6—response A

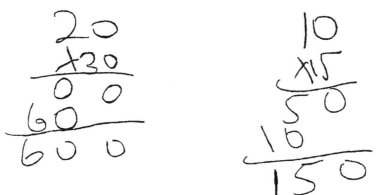

3   E

Mr. Goldstein wants to buy carpet to cover the floors completely in his living room and game room. His living room is 10 feet by 15 feet. His game room is 20 feet by 30 feet.

Mr. Goldstein thinks that the area to be carpeted in his game room is <u>double</u> the area to be carpeted in his living room.

Is Mr. Goldstein correct?   no

    Explain your answer and show your work. You may use diagrams as part of your explanation and work.

Fig. 3.30. Activity 6—response E

3   **F**

Mr. Goldstein wants to buy carpet to cover the floors completely in his living room and game room. His living room is 10 feet by 15 feet. His game room is 20 feet by 30 feet.

Mr. Goldstein thinks that the area to be carpeted in his game room is <u>double</u> the area to be carpeted in his living room.

Is Mr. Goldstein correct?

> Explain your answer and show your work. You may use diagrams as part of your explanation and work.

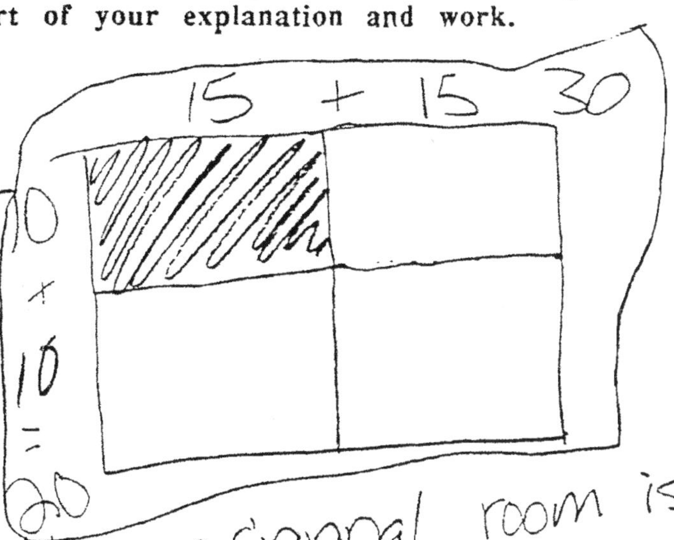

It's actually 3 times the size of his living room as you see in my scketch the original room is shado in but theres still three parts that need to be fill. It seems like half because half of 20 is 10 and half of 30 is 15. But it's really thee times the first room.

Fig.3.31. Activity 6—response F

Mr. Goldstein wants to buy carpet to cover the floors completely in his living room and game room. His living room is 10 feet by 15 feet. His game room is 20 feet by 30 feet.

Mr. Goldstein thinks that the area to be carpeted in his game room is <u>double</u> the area to be carpeted in his living room.

Is Mr. Goldstein correct? *Yes*

    Explain your answer and show your work. You may use diagrams as part of your explanation and work.

See the living room is 10 feet by 15 feet + his game room is 20 feet by 30 feet + 20 is double 10 + 30 is double 15.

$$\begin{array}{r} 20 \\ -10 \\ \hline 10 \end{array} \qquad \begin{array}{r} 30 \\ -15 \\ \hline 15 \end{array}$$

Fig.3.32. Activity 6—response M

CHAPTER 4

THE TASK PACKETS

BLOCKS

Blocks Task

Mathematical Content

Numbers and operations

Task Description

This task allows students to demonstrate their understanding of number sense and their problem-solving abilities by using basic concepts of number theory. Students are asked to find the total number of blocks that when placed in groups of two, three, and four result in one block remaining. Two constraints must be satisfied: (1) the same total number of blocks is partitioned each time into groups of two, three, and four; and (2) the total number of blocks has a remainder of 1 when divided by 2, 3, and 4. Students are asked to provide an answer and show their work. A number of strategies can be used to solve the problem, such as finding common multiples, interpreting remainders in division computations, and using diagrams to represent groups of blocks. Multiple correct answers are also possible for this task, including 13, 25, 37, and so on.

Parke, Carol S., Suzanne Lane, Edward A. Silver, and Maria E. Magone. *Using Assessment to Improve Middle-Grades Mathematics Teaching and Learning: Suggested Activities Using QUASAR Tasks, Scoring Criteria, and Students' Work.* Reston, Va.: National Council of Teachers of Mathematics, 2003.

Blocks Task

Yolanda was telling her brother Damian about what she did in math class.

Yolanda said, "Damian, I used blocks in my math class today. When I grouped the blocks in groups of 2, I had 1 block left over. When I grouped the blocks in groups of 3, I had 1 block left over. When I grouped the blocks in groups of 4, I still had 1 block left over."

Damian asked, "How many blocks did you have?"

What was Yolanda's answer to her brother's question?

Show your work.

Answer: _____

BLOCKS TASK SCORING CRITERIA

Level 4

Work, diagram, or explanation shows a correct and complete understanding of common multiples and of the conditions of the task. The two conditions in the task are (1) the same total number of blocks is partitioned each time into groups of two, three, and four; and (2) the correct numerical answer has a remainder of 1 when divided by 2, 3, and 4.

Level 3

Work, diagram, or explanation shows a nearly correct and complete understanding of common multiples and of the two conditions of the task. However, the response contains a minor error or omission, such as an incorrect listing of common multiples for one of the numbers.

Level 2

Work, diagram, or explanation shows some understanding of multiples and of the conditions of the task. However, one or both of the conditions may not be satisfied for all groupings of the blocks. For instance, the condition of the remainder of 1 is satisfied for only two of the three groupings of blocks.

Level 1

Work, diagram, or explanation shows a limited understanding of multiples or of the conditions of the task. Only one group of two, three, or four blocks with a remainder is shown, or only a correct answer is given with no work.

Level 0

No understanding of multiples or the conditions of the task is shown.

Rationales for Scored Student Responses to Blocks Task

Label	Score	Rationale
A	4	Correct answer (13). Work shows a listing of multiples for 2, 3, and 4. The explanation identifies 12 as the first common multiple and indicates that the remainder of 1 should be added to produce the final answer.
B	4	Correct answer (13). Diagrams show groupings of blocks by 2, 3, and 4. The total number of blocks in each diagram is a constant, and each diagram shows 1 block remaining.
C	4	Correct answer (13). Work shows mathematical computations that produce a common multiple for 2, 3, and 4. Then 1 more block is added to represent the block that is left over.
D	4	Multiple correct answers (13, 25, …). Work shows a listing of multiples for groups of 2, 3, and 4. Several common multiples are found, and the explanation indicates how to obtain several possible totals for the number of blocks.
E	3	Correct answer (13). The student makes some errors in representing the problem. Instead of placing the blocks in groups of 2, 3, and 4, the student divides them into two groups, three groups, and four groups.
F	3	Correct answer (13). The correct multiple of 12 is found for 2, 3, and 4. A remainder of 1 is found, but the division is shown incorrectly, that is, $12 \div 13 = 1$.
G	3	Incorrect answer. Work shows a listing of multiples for 2, 3, and 4. A common multiple is found, and 1 is added to represent the remainder. However, an error in the list of multiples for 3 leads to an incorrect answer.
H	2	Incorrect answer. Diagrams show that the total number of blocks (19) is kept constant for all three groupings. However, the condition of the remainder of 1 is satisfied only for the groups of 2 and 3 but not for the group of 4.
I	2	Incorrect answer. Correct lists of multiples for 2, 3, and 4 are shown; however, the method for obtaining the answer is unclear.
J	2	Incorrect answer. Multiple groupings of 2, 3, and 4 blocks are shown, each with a remainder of 1. The total number of blocks in each grouping, however, is not the same. The student then finds the total number of blocks for all groupings.
K	1	Correct answer (13). No work or diagrams are shown. Explanation only restates the answer.
L	1	Incorrect answer. Diagrams show only one group of 2, one group of 3, and one group of 4, with a remainder of 1 represented for each. The answer of 12 is the sum of the number of blocks drawn for all groupings.
M	0	Incorrect answer. Work shows a meaningless manipulation of numbers extracted from the task. No understanding is evident.
N	0	No answer. No work or diagrams are provided. Explanation indicates no understanding of multiples or of the conditions of the task.
P	0	Incorrect answer. No work shown or explanation given.

A 4

Yolanda was telling her brother Damian about what she did in math class.

Yolanda said, "Damian, I used blocks in my math class today. When I grouped the blocks in groups of 2, I had 1 block left over. When I grouped the blocks in groups of 3, I had 1 block left over. And when I grouped the blocks in groups of 4, I still had 1 block left over."

Damian asked, "How many blocks did you have?"

What was Yolanda's answer to her brother's question?

Show your work. Well first I find a comon multiple of all the

2, 4, 6, 8, 10, (12), 14, 16, 8
3, 6, 9, (12), 15, 18
4, 8, (12), 16

12 is the first comon so I add 1 and my awnsers

→ 13 ←

Answer: 13

B

4

Yolanda was telling her brother Damian about what she did in math class.

Yolanda said, "Damian, I used blocks in my math class today. When I grouped the blocks in groups of 2, I had 1 block left over. When I grouped the blocks in groups of 3, I had 1 block left over. And when I grouped the blocks in groups of 4, I still had 1 block left over."

Damian asked, "How many blocks did you have?"

What was Yolanda's answer to her brother's question?

Show your work.

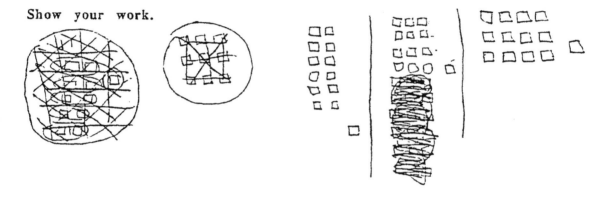

Answer: ___13 blocks___

C   4

Yolanda was telling her brother Damian about what she did in math class.

Yolanda said, "Damian, I used blocks in my math class today. When I grouped the blocks in groups of 2, I had 1 block left over. When I grouped the blocks in groups of 3, I had 1 block left over. And when I grouped the blocks in groups of 4, I still had 1 block left over."

Damian asked, "How many blocks did you have?"

What was Yolanda's answer to her brother's question?

Show your work.

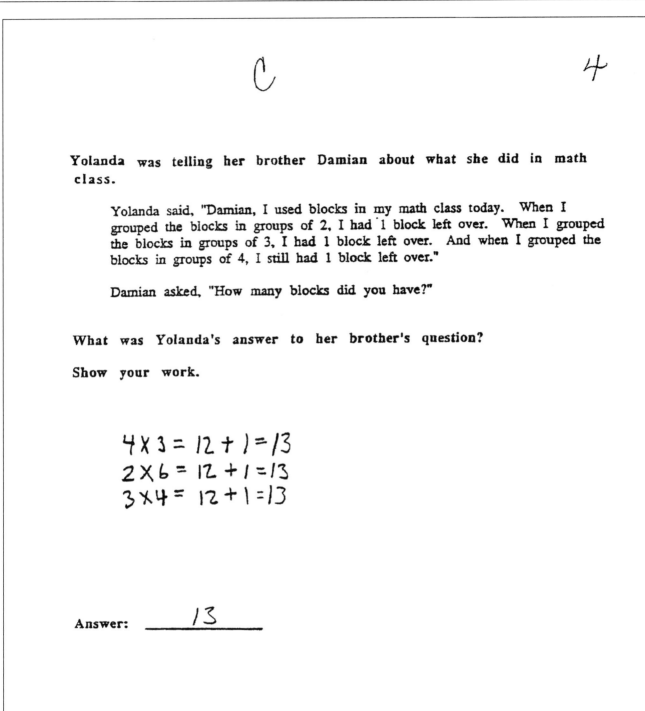

$$4 \times 3 = 12 + 1 = 13$$
$$2 \times 6 = 12 + 1 = 13$$
$$3 \times 4 = 12 + 1 = 13$$

Answer: _____13_____

D   4

Yolanda was telling her brother Damian about what she did in math class.

Yolanda said, "Damian, I used blocks in my math class today. When I grouped the blocks in groups of 2, I had 1 block left over. When I grouped the blocks in groups of 3, I had 1 block left over. And when I grouped the blocks in groups of 4, I still had 1 block left over."

Damian asked, "How many blocks did you have?"

What was Yolanda's answer to her brother's question?

Show your work.

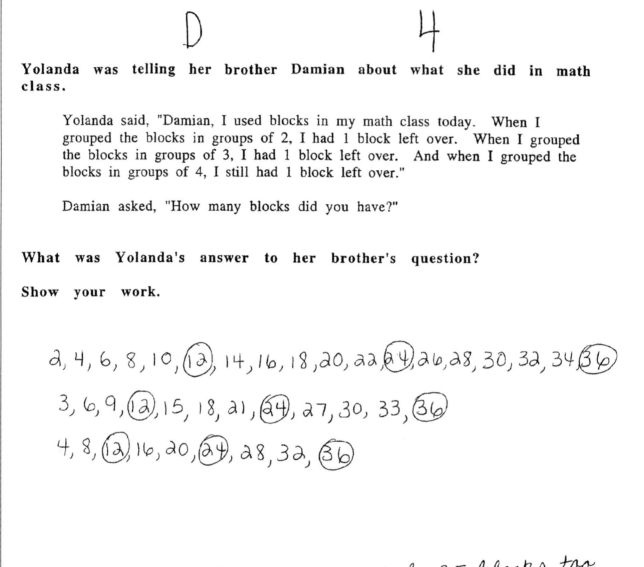

Answer: <u>13 blocks,</u> but there could be 25 blocks too, and it can go on like 12, 24, 36, 48, ...

and add 1 to each of them for the left over block to get

13, 25, 37, 49, ...

E   3

Yolanda was telling her brother Damian about what she did in math class.

Yolanda said, "Damian, I used blocks in my math class today. When I grouped the blocks in groups of 2, I had 1 block left over. When I grouped the blocks in groups of 3, I had 1 block left over. And when I grouped the blocks in groups of 4, I still had 1 block left over."

Damian asked, "How many blocks did you have?"

What was Yolanda's answer to her brother's question?

Show your work.

Groups of two

◻ Left over

Groups of 3

◻ leftover

Groups of 4

◻ left

Answer: _____13_____

F   3

Yolanda was telling her brother Damian about what she did in math class.

Yolanda said, "Damian, I used blocks in my math class today. When I grouped the blocks in groups of 2, I had 1 block left over. When I grouped the blocks in groups of 3, I had 1 block left over. And when I grouped the blocks in groups of 4, I still had 1 block left over."

Damian asked, "How many blocks did you have?"

What was Yolanda's answer to her brother's question?

Show your work.

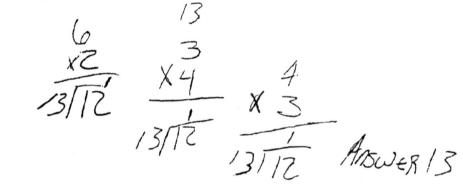

Answer: __13__

G   3

Yolanda was telling her brother Damian about what she did in math class.

Yolanda said, "Damian, I used blocks in my math class today. When I grouped the blocks in groups of 2, I had 1 block left over. When I grouped the blocks in groups of 3, I had 1 block left over. And when I grouped the blocks in groups of 4, I still had 1 block left over."

Damian asked, "How many blocks did you have?"

What was Yolanda's answer to her brother's question?

Show your work.

she had an odd number of blocks

2 4 6 8 10 12 14 16 18 20 22 24 26 28 30 32 34 36 38 40 42 44

3 6 9 12 15 18 21 23 26 27 30 33 36 39 41 44

4 8 12 16 20 24 28 32 36 40 44

Answer: __45 blocks__

H   2

Yolanda was telling her brother Damian about what she did in math class.

Yolanda said, "Damian, I used blocks in my math class today. When I grouped the blocks in groups of 2, I had 1 block left over. When I grouped the blocks in groups of 3, I had 1 block left over. And when I grouped the blocks in groups of 4, I still had 1 block left over."

Damian asked, "How many blocks did you have?"

What was Yolanda's answer to her brother's question?

Show your work.

Groups of 2 →

Groups of 3 →

Answer: 19

I   2

Yolanda was telling her brother Damian about what she did in math class.

Yolanda said, "Damian, I used blocks in my math class today. When I grouped the blocks in groups of 2, I had 1 block left over. When I grouped the blocks in groups of 3, I had 1 block left over. And when I grouped the blocks in groups of 4, I still had 1 block left over."

Damian asked, "How many blocks did you have?"

What was Yolanda's answer to her brother's question?

Show your work.

2, 4, 6, 8, 10, 12, 14, 16, 18, 20.
3, 6, 9, 12, 15, 18, 21, 24, 27
4, 8, 12, 16, 20, 24, 28, 32
1, 5, 7, 11

Answer: ___5___

J 2

Yolanda was telling her brother Damian about what she did in math class.

Yolanda said, "Damian, I used blocks in my math class today. When I grouped the blocks in groups of 2, I had 1 block left over. When I grouped the blocks in groups of 3, I had 1 block left over. And when I grouped the blocks in groups of 4, I still had 1 block left over."

Damian asked, "How many blocks did you have?"

What was Yolanda's answer to her brother's question?

Show your work.

Answer: She has 21 Blocks in all

K

1

Yolanda was telling her brother Damian about what she did in math class.

Yolanda said, "Damian, I used blocks in my math class today. When I grouped the blocks in groups of 2, I had 1 block left over. When I grouped the blocks in groups of 3, I had 1 block left over. And when I grouped the blocks in groups of 4, I still had 1 block left over."

Damian asked, "How many blocks did you have?"

What was Yolanda's answer to her brother's question?

Show your work.

She had 13 block all together

Answer: ___13___

Yolanda was telling her brother Damian about what she did in math class.

Yolanda said, "Damian, I used blocks in my math class today. When I grouped the blocks in groups of 2, I had 1 block left over. When I grouped the blocks in groups of 3, I had 1 block left over. And when I grouped the blocks in groups of 4, I still had 1 block left over."

Damian asked, "How many blocks did you have?"

What was Yolanda's answer to her brother's question?

Show your work.

$$\begin{array}{r} 2 \\ 3 \\ +\ 4 \\ \hline 9 \end{array}$$

Answer: ___12___

M

Yolanda was telling her brother Damian about what she did in math class.

Yolanda said, "Damian, I used blocks in my math class today. When I grouped the blocks in groups of 2, I had 1 block left over. When I grouped the blocks in groups of 3, I had 1 block left over. And when I grouped the blocks in groups of 4, I still had 1 block left over."

Damian asked, "How many blocks did you have?"

What was Yolanda's answer to her brother's question?

Show your work.

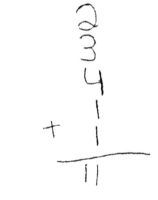

Answer: __11__

N

Yolanda was telling her brother Damian about what she did in math class.

Yolanda said, "Damian, I used blocks in my math class today. When I grouped the blocks in groups of 2, I had 1 block left over. When I grouped the blocks in groups of 3, I had 1 block left over. And when I grouped the blocks in groups of 4, I still had 1 block left over."

Damian asked, "How many blocks did you have?"

What was Yolanda's answer to her brother's question?

Show your work.

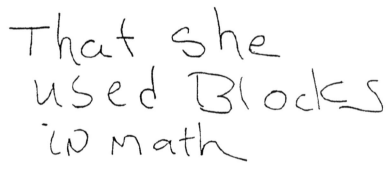

Answer: _____

P   O

Yolanda was telling her brother Damian about what she did in math class.

Yolanda said, "Damian, I used blocks in my math class today. When I grouped the blocks in groups of 2, I had 1 block left over. When I grouped the blocks in groups of 3, I had 1 block left over. And when I grouped the blocks in groups of 4, I still had 1 block left over."

Damian asked, "How many blocks did you have?"

What was Yolanda's answer to her brother's question?

Show your work.

Answer: _____9 blocks_____

Miguel's Homework Task

Mathematical Content

Algebra, functions, and patterns

Task Description

In this pattern task, students demonstrate the ability to recognize the underlying mathematical structure used to generate a visual pattern, then describe the pattern. Specifically, students are asked to draw the fifth figure in a pattern given the first four figures. Students must also describe how they know which figure comes next in the pattern. This visual pattern involves two regularities: (1) the number of dots added to each figure and (2) the shape of the figure or the positions in the pattern where the dots are to be added. The students' drawings and descriptions of the pattern must include both of these regular features; that is, (1) three dots must be added to each subsequent figure, and (2) one dot must be added to each row of the figure to get the next figure.

Parke, Carol S., Suzanne Lane, Edward A. Silver, and Maria E. Magone. *Using Assessment to Improve Middle-Grades Mathematics Teaching and Learning: Suggested Activities Using QUASAR Tasks, Scoring Criteria, and Students' Work.* Reston, Va.: National Council of Teachers of Mathematics, 2003.

Miguel's Homework Task

For homework Miguel's teacher asked him to look at the pattern below and draw the figure that should come next.

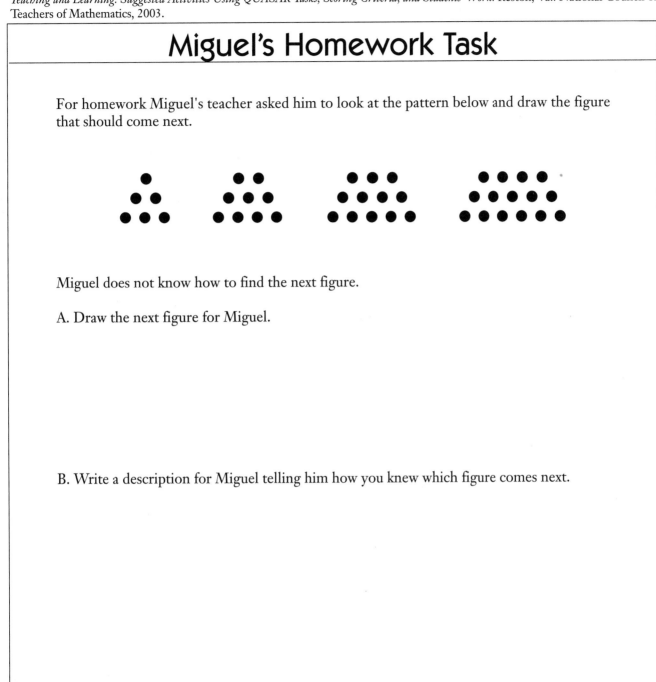

Miguel does not know how to find the next figure.

A. Draw the next figure for Miguel.

B. Write a description for Miguel telling him how you knew which figure comes next.

Miguel's Homework Task Scoring Criteria

Level 4

Both the description and the figure show evidence that the student has detected the regularities in the pattern and described them completely and correctly. The two regularities in this pattern are related to the number of dots and the shape of the figure. Three dots are added to obtain the subsequent figure, one in each row of the preceding figure.

Level 3

Both the description and the figure show evidence that the student has detected and described the regularities in the pattern nearly completely and correctly. Either the figure or the description is slightly incorrect or incomplete. For example, the figure could be correct, but the description states only that the pattern goes up by three and does not describe where the dots are added to the figure.

Level 2

The description or figure shows evidence that the student has detected and partially described some of the regularities in the pattern. For example, the shape of the figure is incorrect, but the figure contains the correct number of dots and the description focuses only on the number of dots but not the shape. Another type of response at this level would be that the figure is correct, but no description is given.

Level 1

The figure is incorrect, but the description or a part of the figure shows evidence that the student has limited understanding of the pattern; for example, only one line of the figure is correct.

Level 0

The description and figure show evidence that the student has no understanding of the pattern. Dots may be drawn, but no aspect of the figure or description is correct.

Rationales for Scored Student Responses to Miguel's Homework Task

Label	Score	Rationale
A	4	Correct figure. Explanation completely and correctly states the number of dots added to each figure and indicates that the dots are added to the side of each figure.
B	4	Correct figure. Complete and correct pictorial description of the pattern with labels that show the increasing numbers of dots in each row of each figure.
C	4	Correct figure. Explanation completely and correctly describes the pattern in a horizontal fashion by showing the increase in the number of dots across the first, second, and third row of each figure.
D	3	Incorrect number of dots in the figure, but the shape of the figure is correct. Explanation completely and correctly indicates the number of dots added to each figure and correctly shows where the dots are added to the figure.
E	3	Correct figure. Explanation is incomplete because it does not specifically indicate that one more dot is added to each row of the figure.
F	2	Incorrect figure; the figure contains the correct number of dots but not the correct shape. Explanation partially identifies the pattern by indicating the increase in dots but is somewhat incomplete because it does not indicate where the three dots are to be added to each figure.
G	2	Correct figure. No explanation of the pattern is given, and the numbers shown indicate only the number of dots in the figure that was drawn.
H	2	Correct figure. Explanation does not describe either of the regularities in the pattern; it states only that the dots in the figure follow a pattern.
I	1	Incorrect figure; the figure contains the correct number of dots but not the correct shape. No explanation is provided.
J	1	Incorrect figure, but the numbers of dots in the first row and the third row are correct. No explanation is provided.
K	1	Incorrect figure; the figure contains the correct number of dots but not the correct shape. The explanation is meaningless.
L	0	The fourth figure given in the pattern is simply redrawn. No explanation is given.
M	0	Incorrect figure. Explanation reveals no understanding.

A   4

For homework Miguel's teacher asked him to look at the pattern below and draw the figure that should come next.

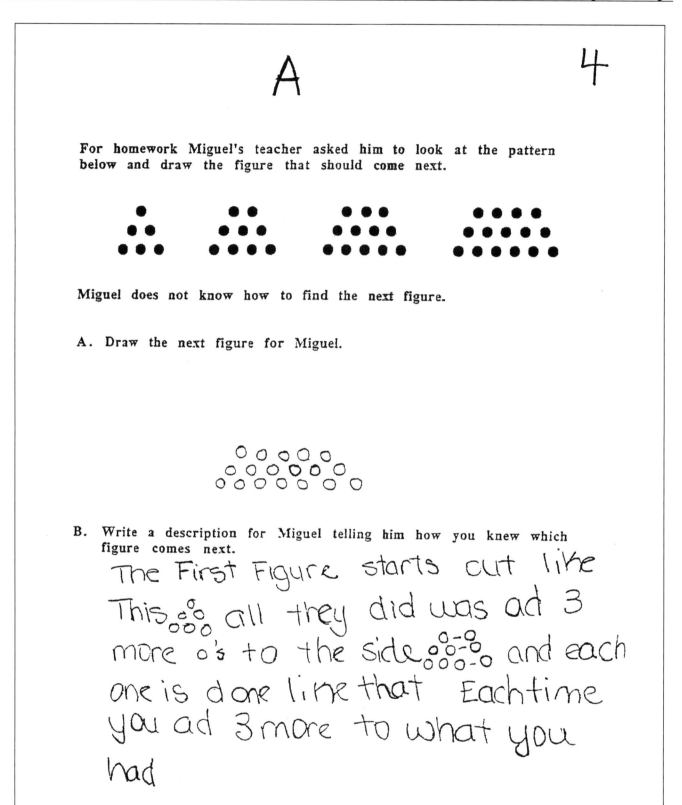

Miguel does not know how to find the next figure.

A. Draw the next figure for Miguel.

B. Write a description for Miguel telling him how you knew which figure comes next.

The First Figure starts out like
This °o° all they did was ad 3
more o's to the side o-o o-o and each
one is done like that Eachtime
you ad 3 more to what you
had

B   4

For homework Miguel's teacher asked him to look at the pattern below and draw the figure that should come next.

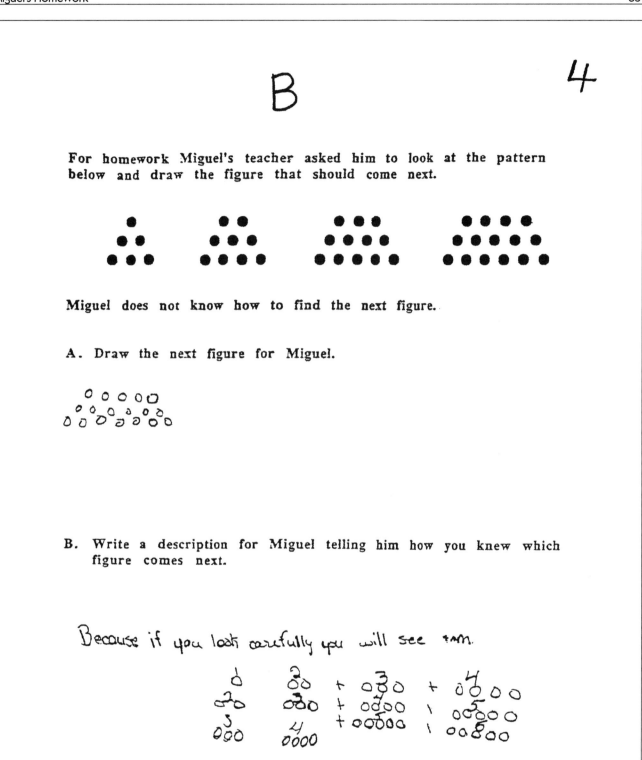

Miguel does not know how to find the next figure.

A. Draw the next figure for Miguel.

B. Write a description for Miguel telling him how you knew which figure comes next.

Because if you look carefully you will see +nm.

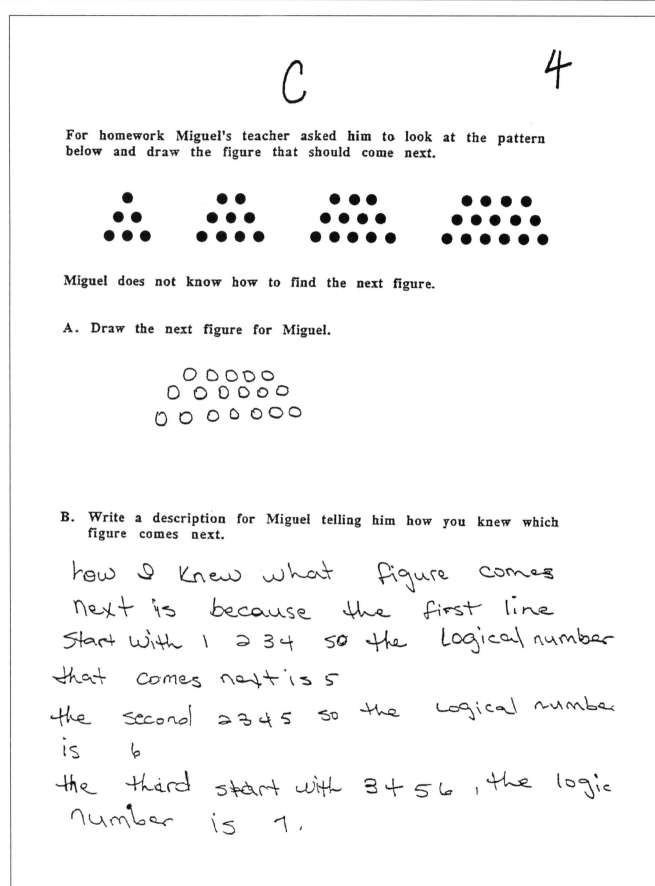

C   4

For homework Miguel's teacher asked him to look at the pattern
below and draw the figure that should come next.

Miguel does not know how to find the next figure.

A. Draw the next figure for Miguel.

B. Write a description for Miguel telling him how you knew which
figure comes next.

how I knew what figure comes
next is because the first line
start with 1 2 3 4 so the logical number
that comes next is 5
the second 2 3 4 5 so the logical number
is 6
the third start with 3 4 5 6, the logic
number is 7.

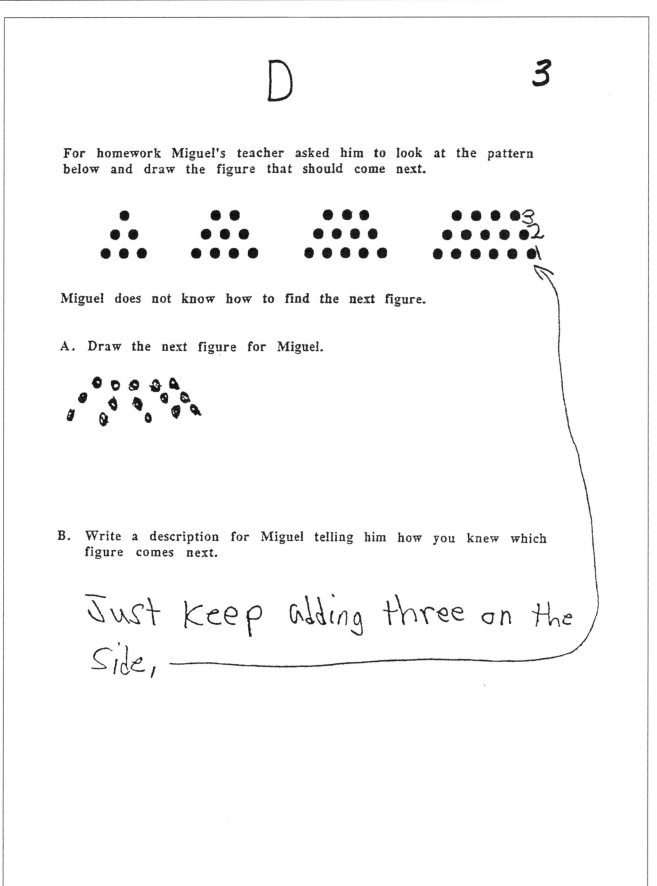

D   3

For homework Miguel's teacher asked him to look at the pattern below and draw the figure that should come next.

Miguel does not know how to find the next figure.

A. Draw the next figure for Miguel.

B. Write a description for Miguel telling him how you knew which figure comes next.

Just keep adding three on the side,

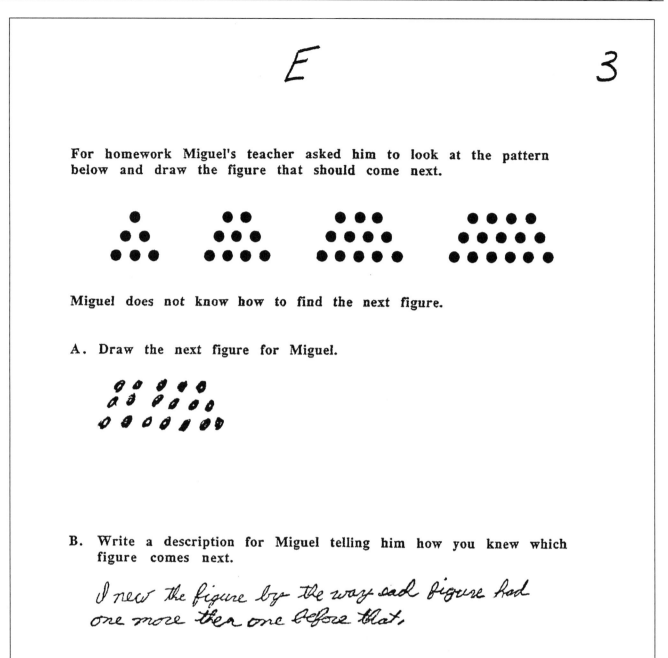

E   *3*

For homework Miguel's teacher asked him to look at the pattern below and draw the figure that should come next.

Miguel does not know how to find the next figure.

A. Draw the next figure for Miguel.

B. Write a description for Miguel telling him how you knew which figure comes next.

I new the figure by the way sad figure had one more then one before that,

F   2

For homework Miguel's teacher asked him to look at the pattern below and draw the figure that should come next.

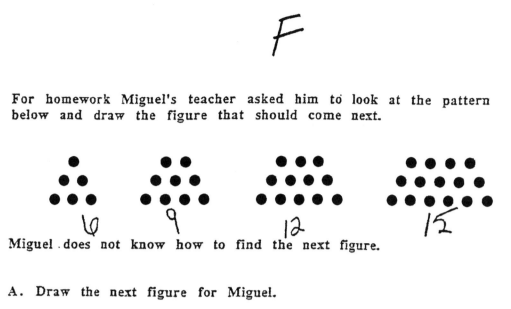

6   9   12   15

Miguel does not know how to find the next figure.

A. Draw the next figure for Miguel.

18

B. Write a description for Miguel telling him how you knew which figure comes next.

Because they go up by three

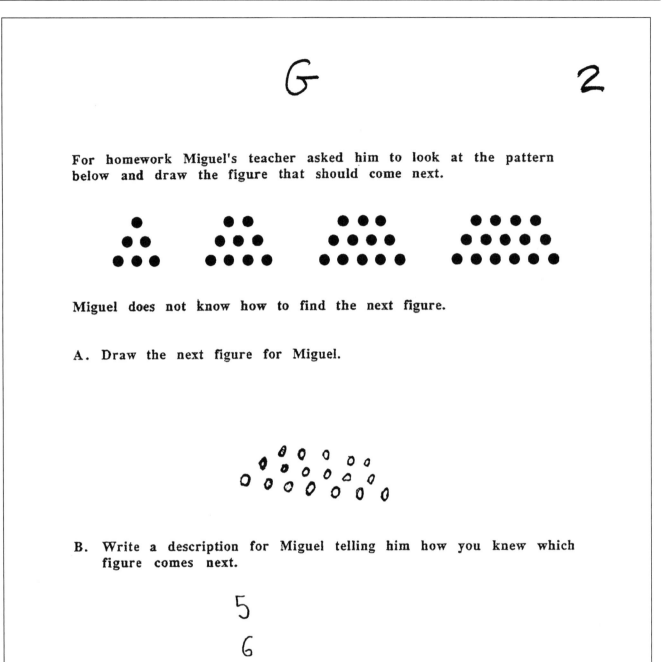

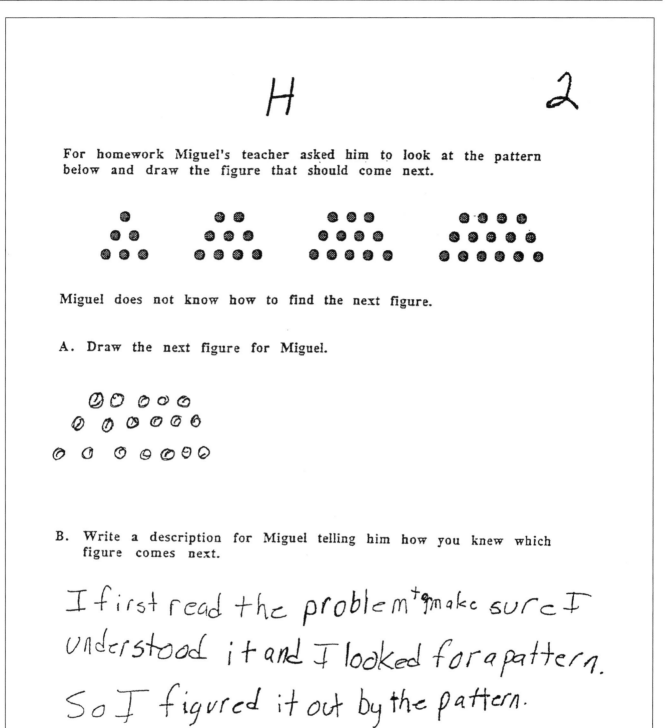

H   2

For homework Miguel's teacher asked him to look at the pattern below and draw the figure that should come next.

Miguel does not know how to find the next figure.

A. Draw the next figure for Miguel.

B. Write a description for Miguel telling him how you knew which figure comes next.

I first read the problem t make sure I understood it and I looked for a pattern. So I figured it out by the pattern.

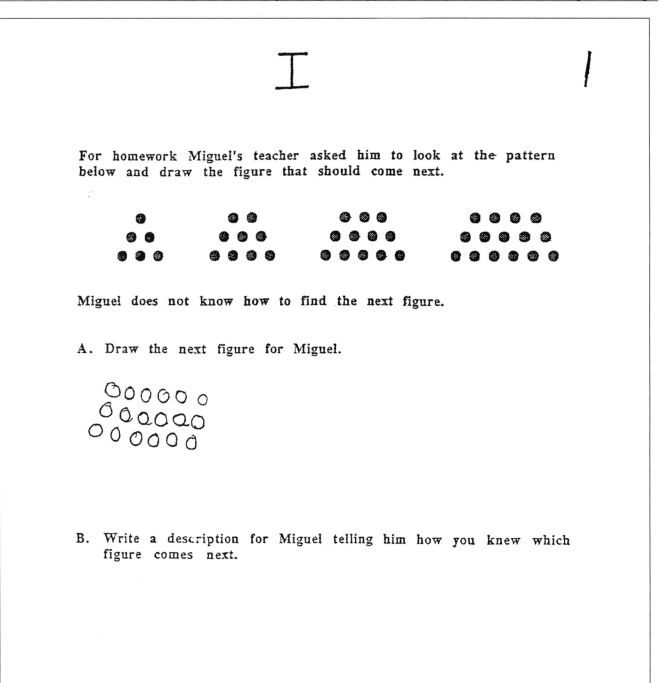

For homework Miguel's teacher asked him to look at the pattern below and draw the figure that should come next.

Miguel does not know how to find the next figure.

A. Draw the next figure for Miguel.

B. Write a description for Miguel telling him how you knew which figure comes next.

J

1

For homework Miguel's teacher asked him to look at the pattern below and draw the figure that should come next.

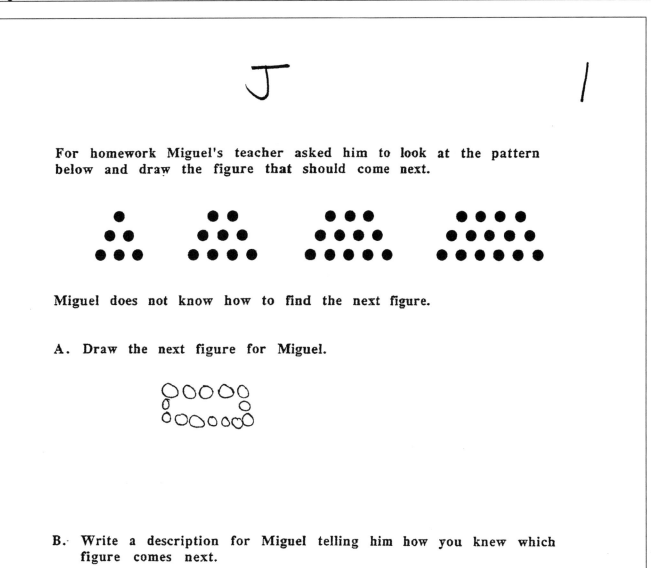

Miguel does not know how to find the next figure.

A. Draw the next figure for Miguel.

B. Write a description for Miguel telling him how you knew which figure comes next.

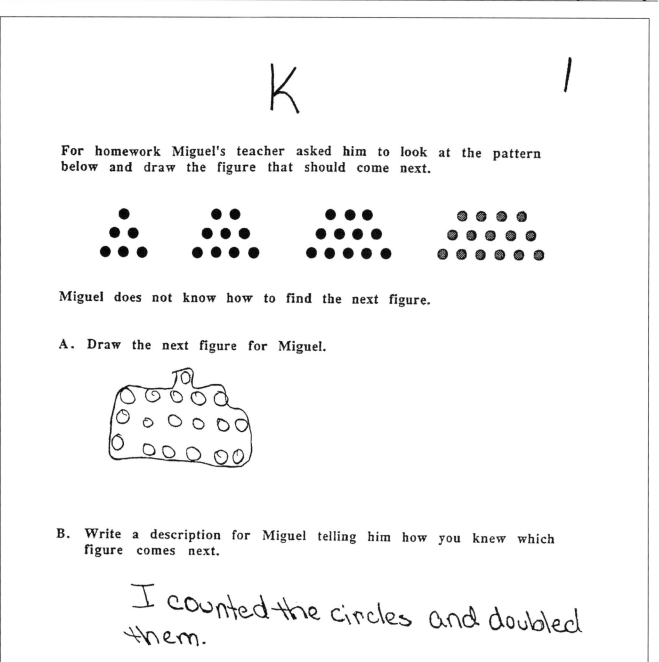

K   1

For homework Miguel's teacher asked him to look at the pattern below and draw the figure that should come next.

Miguel does not know how to find the next figure.

A. Draw the next figure for Miguel.

B. Write a description for Miguel telling him how you knew which figure comes next.

I counted the circles and doubled them.

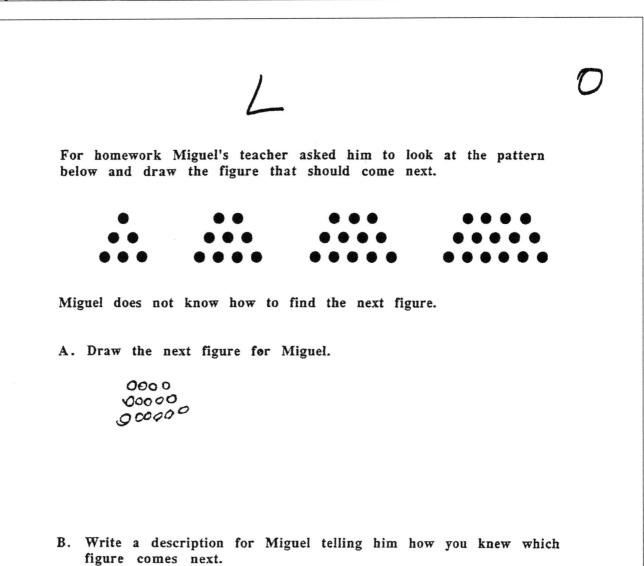

L

For homework Miguel's teacher asked him to look at the pattern below and draw the figure that should come next.

Miguel does not know how to find the next figure.

A. Draw the next figure for Miguel.

B. Write a description for Miguel telling him how you knew which figure comes next.

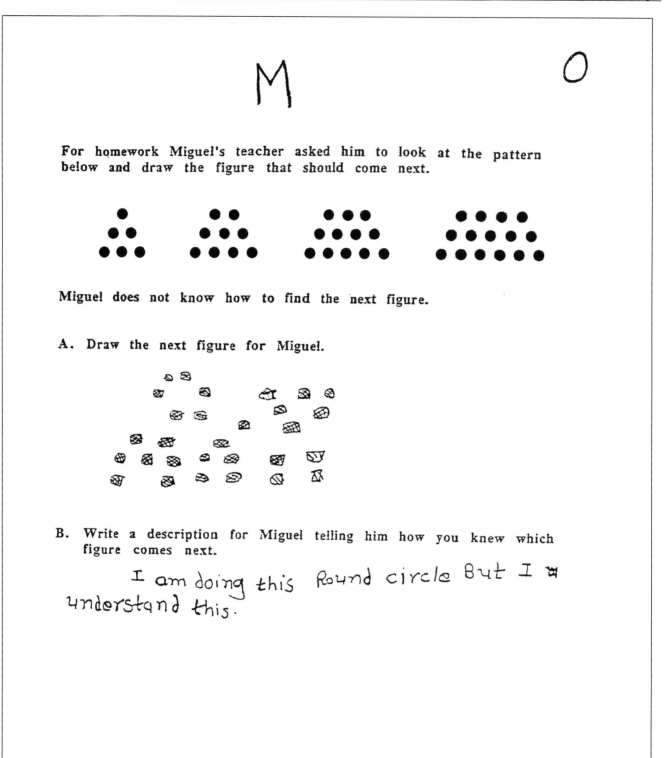

For homework Miguel's teacher asked him to look at the pattern below and draw the figure that should come next.

Miguel does not know how to find the next figure.

A. Draw the next figure for Miguel.

B. Write a description for Miguel telling him how you knew which figure comes next.

I am doing this Round circle But I ⧫ understand this.

BAR AVERAGE

Bar Average Task

Mathematical Content

Data analysis and probability

Task Description

This task focuses on students' understanding of the concept of average. The set of data, represented in a bar graph, consists of scores on three 20-point projects. Students are asked to find the score on the fourth 20-point project when given the average of all four project scores and to explain or show how they found the answer. This task allows for a variety of solution strategies, such as the following:

- Applying the concept of average to determine the total number of points needed across all projects, then working backward to find the answer; for example, $17 \times 4 = 68$, $68 - (15 + 18 + 16) = 19$.

- Substituting x in the average formula for the fourth project, then solving the algebraic equation; that is, $(15 + 18 + 16 + x) \div 4 = 17$, therefore $x = 19$.

- Applying a "leveling" strategy, for example, moving points from one project to another to maintain 17 points for each of the four projects. Students can then determine that the score for the fourth project must be 19 points.

- Guessing scores for the fourth project, then checking each score using the formula for average until the average of 17 is obtained; for example, $15 + 18 + 16 + 19 = 68$, and $68 \div 4 = 17$.

Parke, Carol S., Suzanne Lane, Edward A. Silver, and Maria E. Magone. *Using Assessment to Improve Middle-Grades Mathematics Teaching and Learning: Suggested Activities Using QUASAR Tasks, Scoring Criteria, and Students' Work.* Reston, Va.: National Council of Teachers of Mathematics, 2003.

Bar Average Task

Anita has four 20-point projects for science class. Anita's scores on the first three projects are shown below.

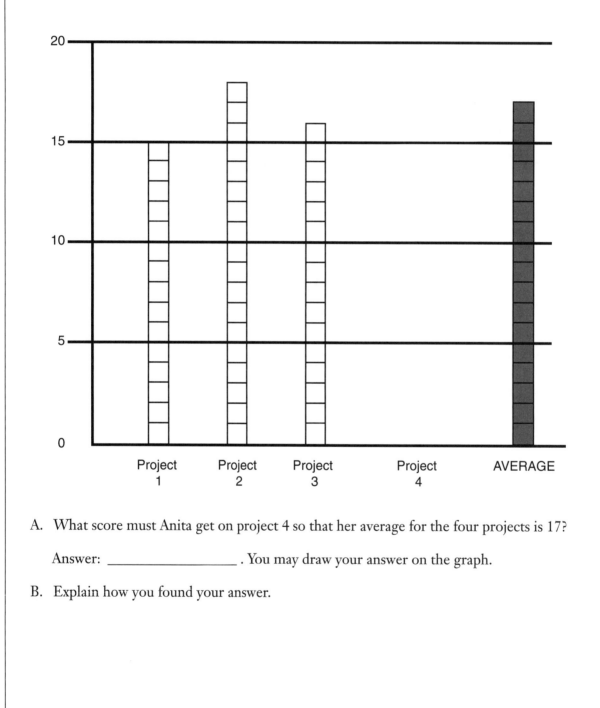

A. What score must Anita get on project 4 so that her average for the four projects is 17?

Answer: _____ . You may draw your answer on the graph.

B. Explain how you found your answer.

Bar Average Task Scoring Criteria

Level 4

Explanation, work, or drawing on the graph shows a correct and complete understanding of the concept of average in the context of the problem. The strategy used to obtain the correct answer is appropriate and is implemented completely and correctly. (Various solution strategies are provided in the task description.)

Level 3

Explanation, work, or drawing on the graph shows a good understanding of the concept of average in the context of the problem; however, the implementation of the strategy contains a minor error or omission. For example, in finding the total of the first three scores, the student may make a calculation error, which leads to an incorrect answer for the fourth score.

Level 2

Explanation, work, or drawing on the graph shows some understanding of the concept of average in the context of the problem, but the use of a strategy to obtain the answer is somewhat incomplete, unclear, or incorrect. For instance, the work may show a correct answer but provide only a general explanation that states that the scores were added to find an answer that worked.

Level 1

A beginning understanding of the concept of average in the context of the problem is revealed. The strategy used to obtain the answer is unclear or incorrect, or no strategy is apparent. The answer may be correct, but no explanation is provided, or possibly, the average of the first three scores is found.

Level 0

No understanding of the concept of average in the context of the problem is evident. Calculations are meaningless, and no explanation is provided or the explanation simply restates the problem.

Rationales for Scored Student Responses to Bar Average Task

Label	Score	Rationale
A	4	Correct answer. The drawing on the graph and the explanation completely and correctly show the use of a "leveling" strategy to remove points from, or add points to, each project to obtain an average of 17 for the four projects.
B	4	Correct answer. Work completely and correctly shows a guess–and-check strategy. Scores for project 4 are selected, then checked to determine which score results in an average of 17 for the four projects.
C	4	Correct answer. The calculations show how the score for project 4 was obtained by using the concept of average. First, the total number of points needed across all four projects was found, then the total points for the first three projects was subtracted from the total points needed.
D	3	Incorrect answer. Work is complete, but a calculation error is made when adding the scores of the four projects. This error leads the student to obtain an average of 17 using an incorrect score for project 4.
E	3	Correct answer. Explanation correctly describes the use of a leveling strategy but is incomplete because it does not explicitly state why 2 more points are needed.
F	3	Incorrect answer. The work is complete and correct given that the score for project 2 was read incorrectly from the graph as 17 instead of 18.
G	2	Incorrect answer. The drawing on the graph shows some understanding of the use of leveling; however, the student makes an error in using the information from the other projects to obtain a score for project 4.
H	2	Incorrect answer. Work shows that a guess-and-check strategy was used, but the answer was not checked correctly because an average of 17 with a remainder is obtained.
I	1	Correct answer. Explanation provides no additional information.
J	1	No answer is given. Work shows a beginning understanding of the concept of average. A guess–and-check strategy is used to some extent, but the total of the four projects is divided by 3 instead of 4.
K	0	Incorrect answer. Explanation states only that "blocks" were counted; no understanding of the average concept is evident.
L	0	Incorrect answer. Work shows no understanding of the average concept.
M	0	Incorrect answer. Explanation shows no understanding of the average concept.

Anita has four 20-point projects for science class.   Anita's scores on the first 3 projects are shown below.

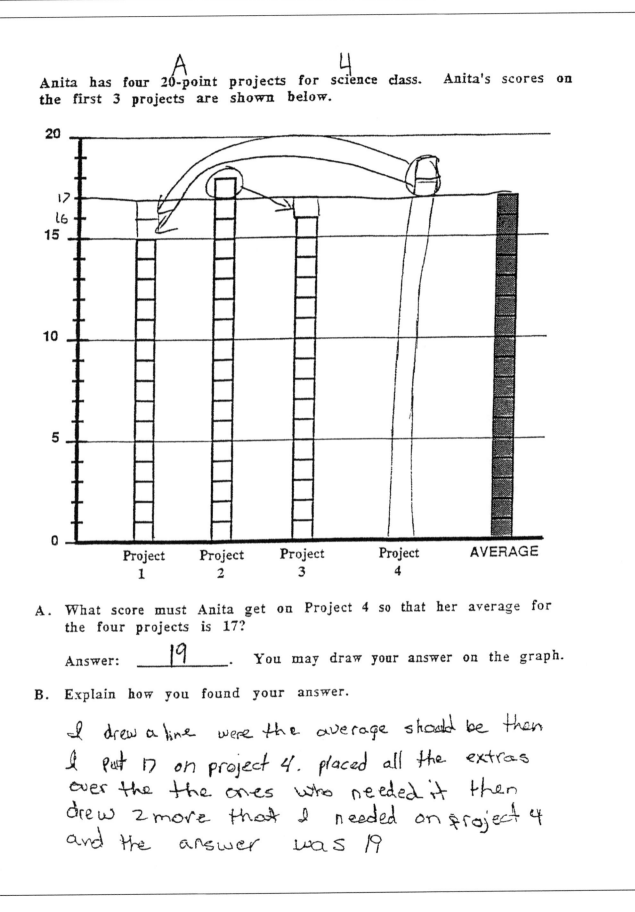

A. What score must Anita get on Project 4 so that her average for the four projects is 17?

Answer: ___19___. You may draw your answer on the graph.

B. Explain how you found your answer.

I drew a line were the average should be then I put 17 on project 4. placed all the extras over the the ones who needed it then drew 2 more that I needed on project 4 and the answer was 19

Anita has four 20-point projects for science class. Anita's scores on the first 3 projects are shown below.

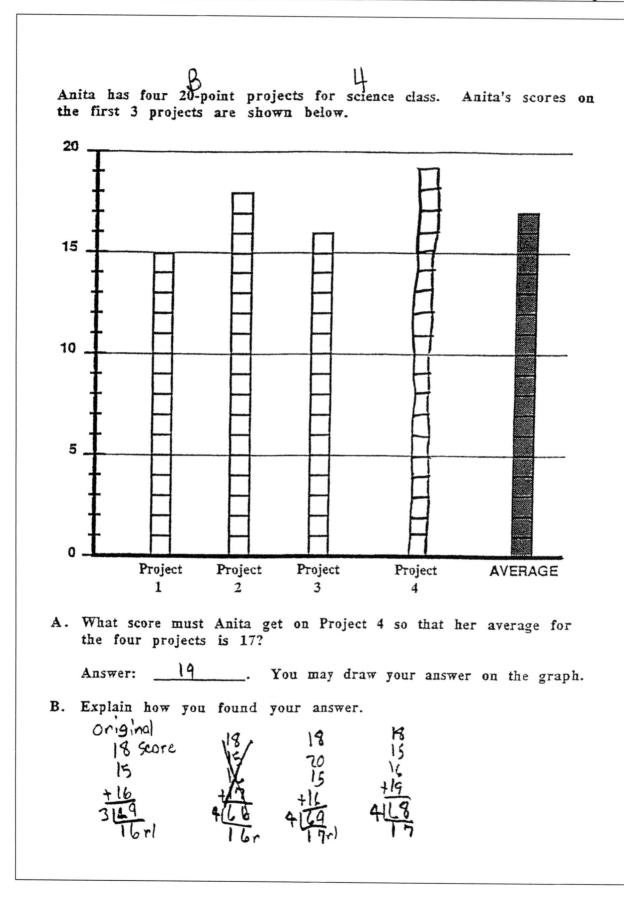

A. What score must Anita get on Project 4 so that her average for the four projects is 17?

Answer: _____19_____. You may draw your answer on the graph.

B. Explain how you found your answer.

C   4

Anita has four 20-point projects for science class.   Anita's scores on
the first 3 projects are shown below.

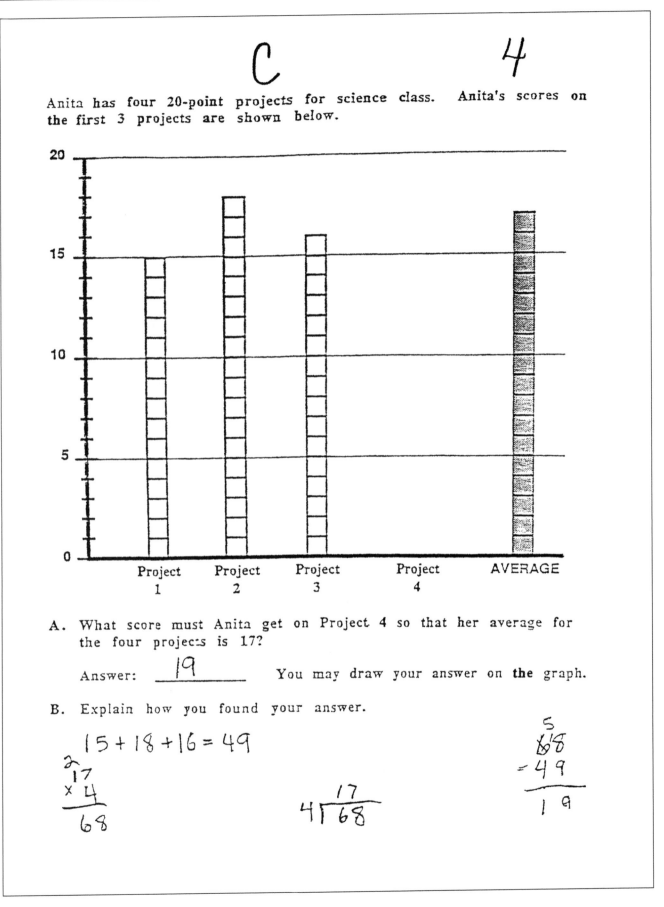

A. What score must Anita get on Project 4 so that her average for
   the four projects is 17?

   Answer: ___19___   You may draw your answer on the graph.

B. Explain how you found your answer.

15 + 18 + 16 = 49

17
x 4
___
68

17
4⟌68

5
68
- 49
___
19

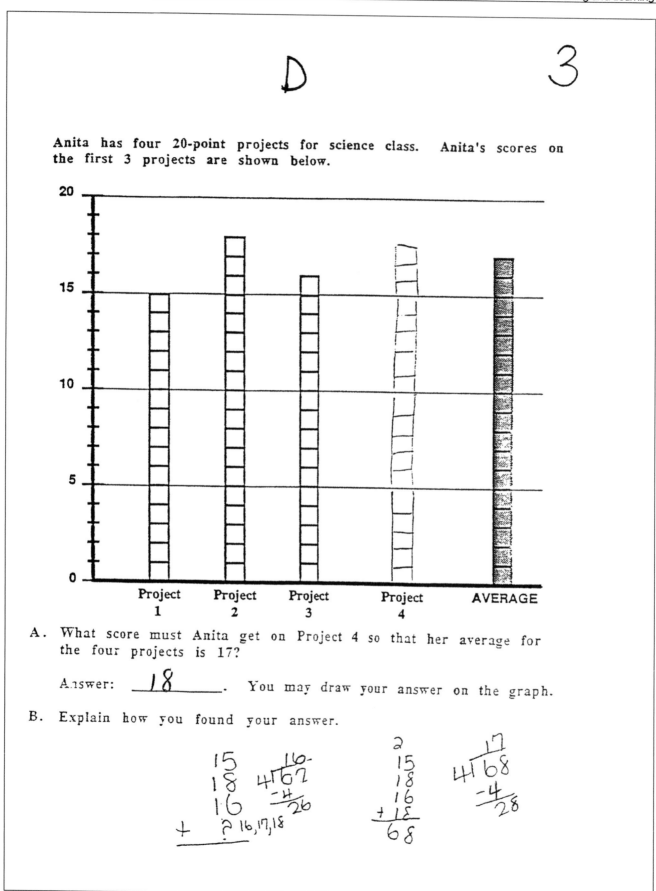

D   3

Anita has four 20-point projects for science class.   Anita's scores on the first 3 projects are shown below.

A. What score must Anita get on Project 4 so that her average for the four projects is 17?

Answer: ___18___.   You may draw your answer on the graph.

B. Explain how you found your answer.

E

3

Anita has four 20-point projects for science class. Anita's scores on the first 3 projects are shown below.

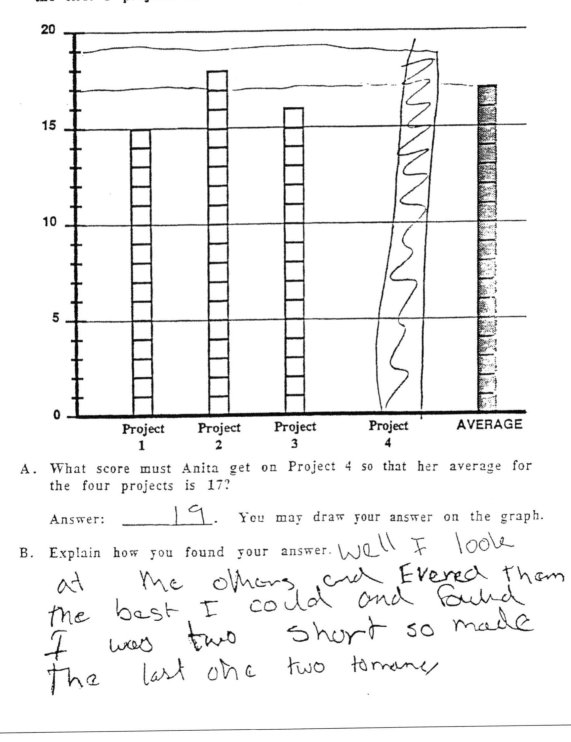

A. What score must Anita get on Project 4 so that her average for the four projects is 17?

Answer: ____19____. You may draw your answer on the graph.

B. Explain how you found your answer. Well I look at the others and Evered them the best I could and found I was two short so made the last one two tomany

F   3

Anita has four 20-point projects for science class. Anita's scores on the first 3 projects are shown below.

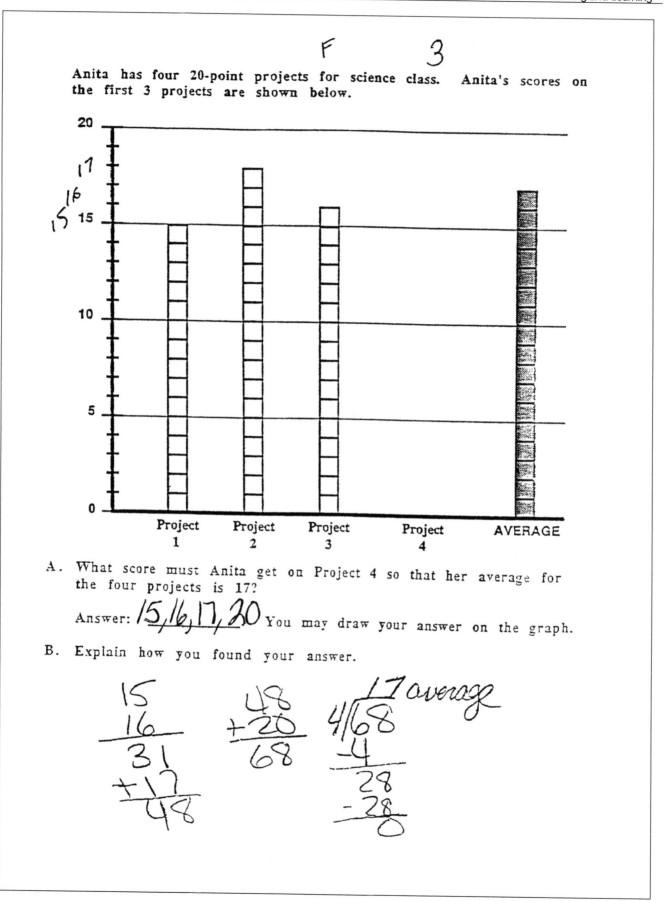

A. What score must Anita get on Project 4 so that her average for the four projects is 17?

Answer: 15, 16, 17, 20 You may draw your answer on the graph.

B. Explain how you found your answer.

G   2

Anita has four 20-point projects for science class.   Anita's scores on the first 3 projects are shown below.

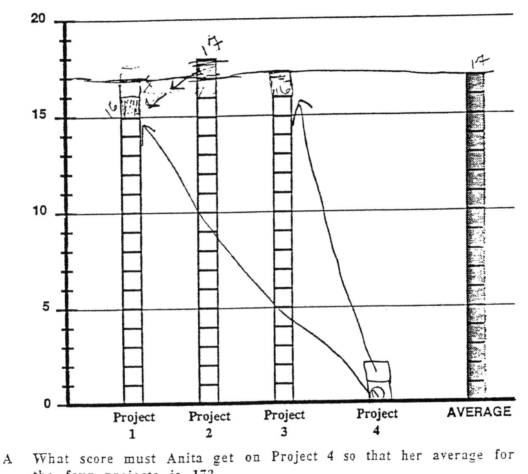

A   What score must Anita get on Project 4 so that her average for the four projects is 17?

Answer: _____2_____.   You may draw your answer on the graph.

B.   Explain how you found your answer.

Because you would have to average out the other scores, and you would have to add 4 points.

Anita has four 20-point projects for H 2 science class. Anita's scores on the first 3 projects are shown below.

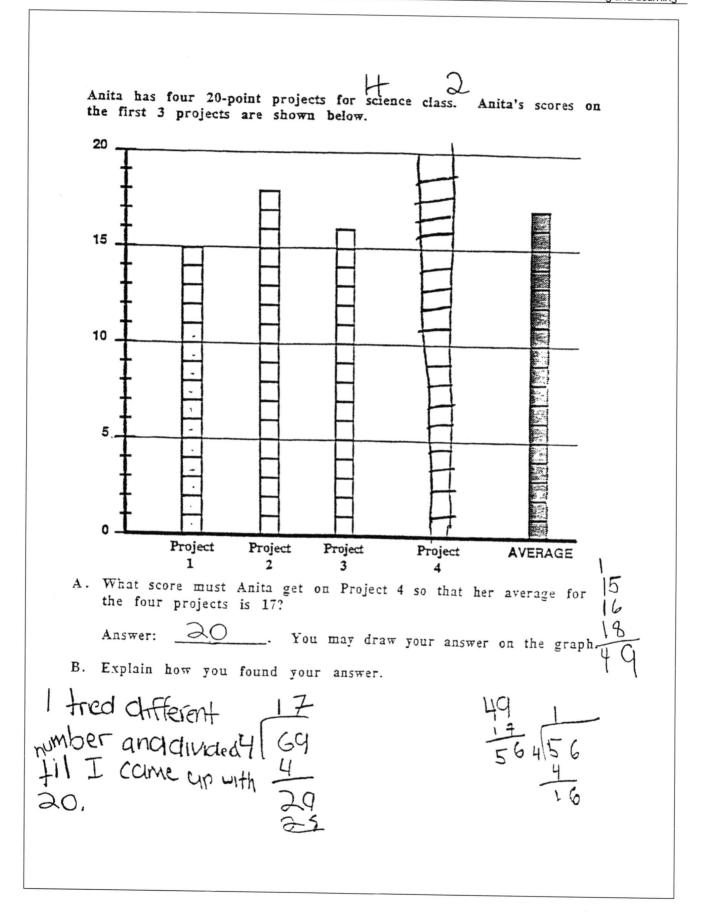

A. What score must Anita get on Project 4 so that her average for the four projects is 17?

Answer: ___20___. You may draw your answer on the graph.

B. Explain how you found your answer.

I tried different number and divided til I came up with 20.

I

Anita has four 20-point projects for science class.   Anita's scores on the first 3 projects are shown below.

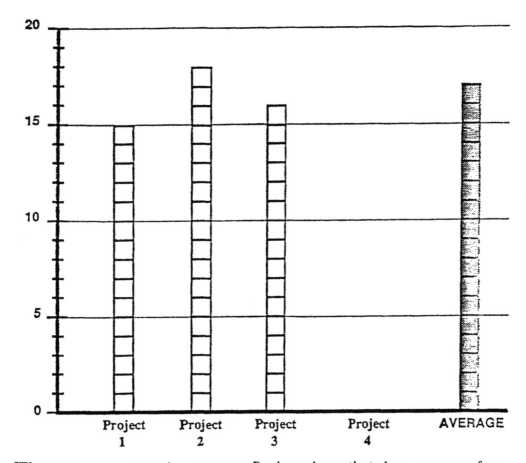

A. What score must Anita get on Project 4 so that her average for the four projects is 17?

Answer: _____19_____.   You may draw your answer on the graph.

B. Explain how you found your answer.

I saw a Paptern

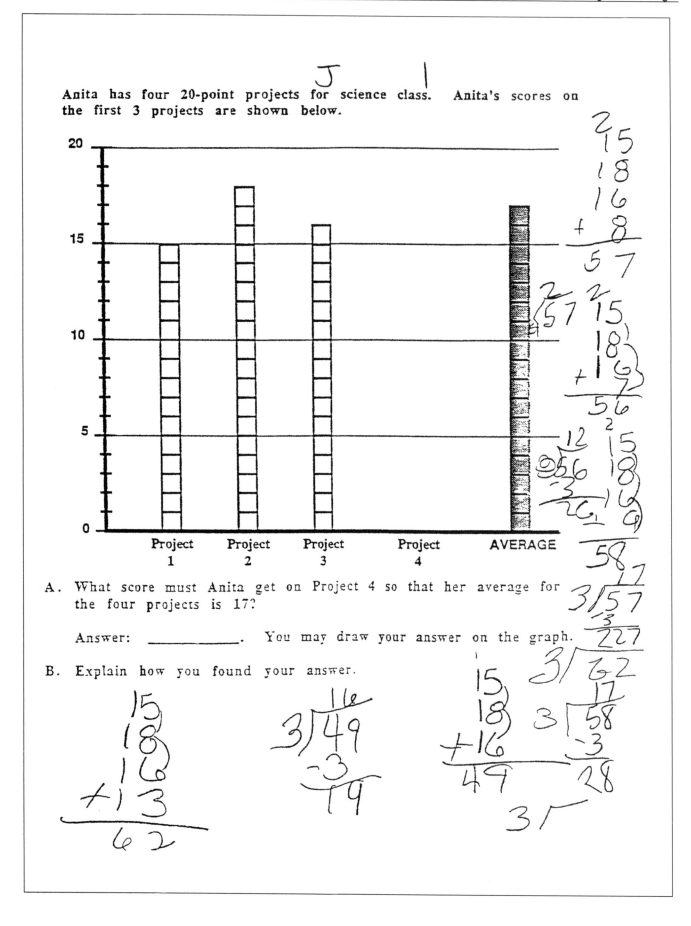

Anita has four 20-point projects for science class.   Anita's scores on the first 3 projects are shown below.

A. What score must Anita get on Project 4 so that her average for the four projects is 17?

Answer: _____.   You may draw your answer on the graph.

B. Explain how you found your answer.

K   O

Anita has four 20-point projects for science class.   Anita's scores on
the first 3 projects are shown below.

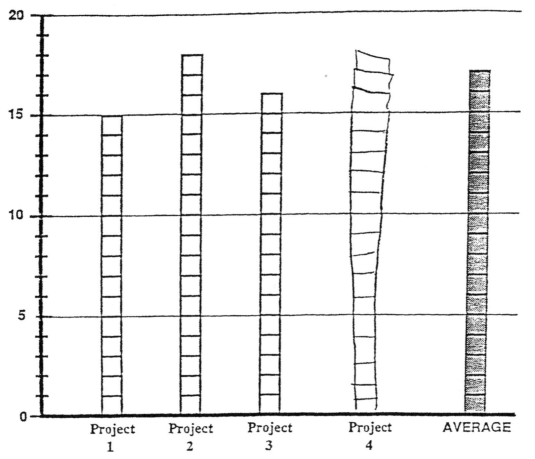

A. What score must Anita get on Project 4 so that her average for
    the four projects is 17?

    Answer: ___18___   You may draw your answer on the graph.

B. Explain how you found your answer.

I You count
the blocks

Anita has four 20-point projects for science class. Anita's scores on the first 3 projects are shown below.

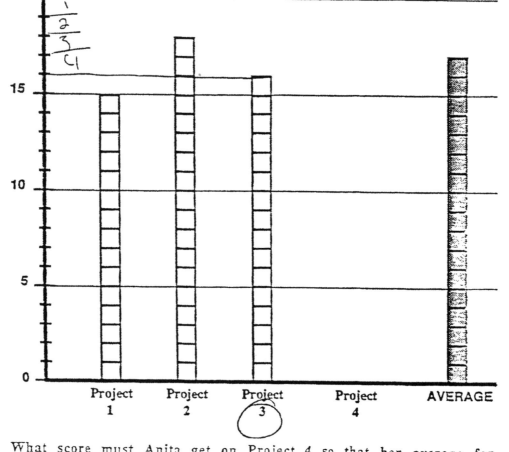

A. What score must Anita get on Project 4 so that her average for the four projects is 17?

Answer: _____. You may draw your answer on the graph.

B. Explain how you found your answer.

M   O

Anita has four 20-point projects for science class. Anita's scores on the first 3 projects are shown below.

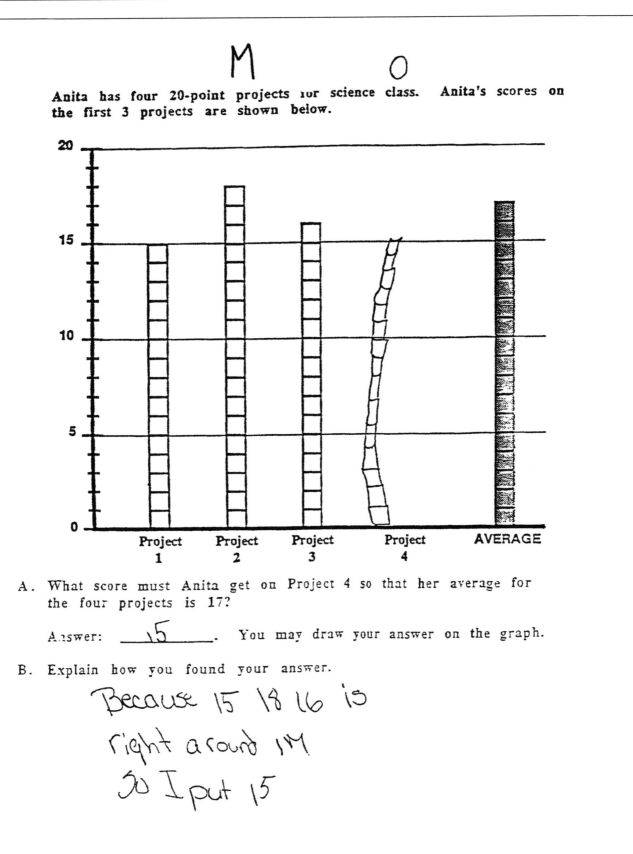

A. What score must Anita get on Project 4 so that her average for the four projects is 17?

Answer: ____15____. You may draw your answer on the graph.

B. Explain how you found your answer.

Because 15 18 16 is
right around 17
so I put 15

SCHOOL BOARD

School Board Task

Mathematical Content

Measurement and geometry

Task Description

The purpose of this task is to determine whether students can differentiate between the concepts of area and perimeter and use them appropriately in two problem situations. The terms area and perimeter do not appear in the task. Instead, the task asks students to demonstrate that they can apply the appropriate measurement when given particular situations. Specifically, students are shown figures of two rectangular plots of land with different dimensions, and they must identify which figure has the larger area, that is, would enclose "as much land as possible," and which figure has the smaller perimeter, that is, would require "less fencing." Students are also asked to show how they found their answers; doing so usually includes showing calculations and comparisons of areas and perimeters of the figures.

Parke, Carol S., Suzanne Lane, Edward A. Silver, and Maria E. Magone. *Using Assessment to Improve Middle-Grades Mathematics Teaching and Learning: Suggested Activities Using QUASAR Tasks, Scoring Criteria, and Students' Work.* Reston, Va.: National Council of Teachers of Mathematics, 2003.

School Board Task

The school board wants to buy a piece of land. Mrs. Gomez and Mr. Langer are each selling a rectangular piece of land next to the school.

Mrs. Gomez's land   **Mr. Langer's land**

A. If the school board wants <u>as much land</u> as possible, should the school board buy from Mrs. Gomez or Mr. Langer?

Show how you found your answer.

Answer: _____

B. If the school board wants to surround either Mrs. Gomez's land or Mr. Langer's land with a fence, which piece of land would require the <u>lesser amount</u> of fencing?

Show how you found your answer.

Answer: _____

SCHOOL BOARD SCORING CRITERIA

Level 4

Correct answers are given, that is, Mr. Langer's land is chosen for parts A and B. Complete and correct work displays a clear understanding of both area and perimeter for solving the problem. In part A, complete and correct calculations are provided for the areas of both pieces of land. In part B, complete and correct calculations are provided for the perimeters of both pieces of land or for the sums of two adjacent sides of both pieces of land.

Level 3

Answers and work display a nearly complete and correct use of both area and perimeter in the two parts of the problem. However, the work contains a minor error in the calculation of areas or perimeters.

Level 2

Answers and work display understanding of either area or perimeter but may not distinguish when to apply each concept. For example, perimeters are calculated for part A and areas are calculated for part B. Another type of response at this level is one that has correct answers but shows no work.

Level 1

Either answers or work shows a limited understanding of some aspect of the problem. The concepts of area, perimeter, or both may be used incorrectly in the solution. For instance, only one area may be calculated, but the calculation is not used in any correct way to obtain an answer.

Level 0

Answers and work display no understanding of the problem nor of the concepts of area and perimeter.

Rationales for Scored Student Responses to School Board Task

Label	Score	Rationale
A	4	Correct answers in parts A and B. Work is complete and correct in both parts, showing calculations for areas in part A and perimeters in part B.
B	4	Correct answers in parts A and B. Work is complete and correct in both parts. The two areas are calculated in part A. The sums of two different sides are calculated and compared in part B.
C	4	Correct answers in parts A and B. Complete and correct work shows the calculation of areas in part A and perimeters in part B.
D	3	Answer and work in part A are complete and correct. Answer in part B is incorrect. Work shows a correct method of finding perimeter.
E	3	Work is complete and correct in part A, but the answer is incorrect. Answer and work in part B are complete and correct.
F	3	Complete and correct work and answer in part A. Correct work in part B, but the answer is incorrect.
G	2	Correct answer in part B. Correct addition and comparison of adjacent sides in part B. Incorrect answer in part A, and no work is shown in part A.
H	2	Complete and correct answer in part B with correct work. Work in part A shows calculations of the perimeter instead of the area.
I	2	Correct answer in part A with complete and correct work. Incorrect answer in part B with incomplete work.
J	1	Correct answers in parts A and B, but no work is shown and the explanation adds nothing to the response.
K	1	Incorrect answer in part A. Work in part A shows calculations of the perimeters instead of the areas. Correct answer in part B, but no work is shown.
L	1	Correct answer in part A. Explanation attempts to argue that the shape closer to a square has the larger area and the smaller perimeter. Answer in part B is incorrect.
M	0	Correct answer only in part B. Incorrect answer in part A. No work is shown.
N	0	Work in part A reveals no understanding of area.
P	0	Correct answer only in part B. Answer and work in part A are incorrect.

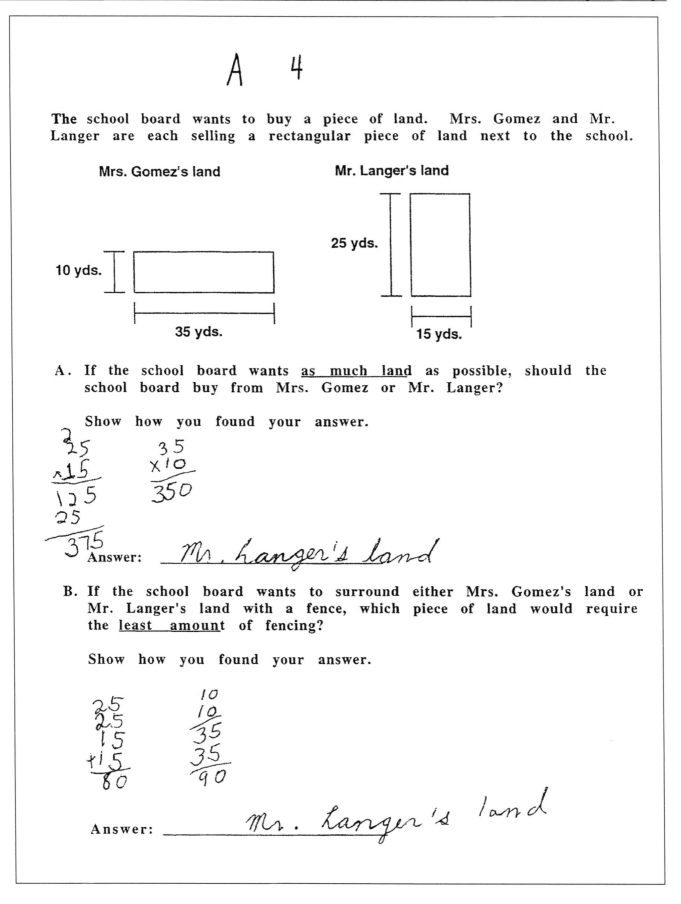

A   4

The school board wants to buy a piece of land. Mrs. Gomez and Mr. Langer are each selling a rectangular piece of land next to the school.

Mrs. Gomez's land

10 yds.

35 yds.

Mr. Langer's land

25 yds.

15 yds.

A. If the school board wants <u>as much land</u> as possible, should the school board buy from Mrs. Gomez or Mr. Langer?

Show how you found your answer.

```
 2
 35        35
x15       x10
125       350
25
375
```

Answer: _Mr. Langer's land_

B. If the school board wants to surround either Mrs. Gomez's land or Mr. Langer's land with a fence, which piece of land would require the <u>least amount</u> of fencing?

Show how you found your answer.

```
 25        10
 25        10
 15        35
+15        35
 80        90
```

Answer: _Mr. Langer's land_

B 4

The school board wants to buy a piece of land. Mrs. Gomez and Mr. Langer are each selling a rectangular piece of land next to the school.

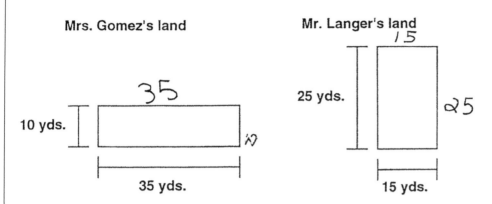

Mrs. Gomez's land

Mr. Langer's land

A. If the school board wants <u>as much land</u> as possible, should the school board buy from Mrs. Gomez or Mr. Langer?

Show how you found your answer.

Gomez 35 × 10 = 350

Langer

$$\begin{array}{r} 25 \\ \underline{15} \\ 125 \\ \underline{250} \\ 375 \end{array}$$

Answer: MR. LangeR

B. If the school board wants to surround either Mrs. Gomez's land or Mr. Langer's land with a fence, which piece of land would require the <u>least amount</u> of fencing?

Show how you found your answer.

MRS Gomez

$$\begin{array}{r} 35 \\ + \ .10 \\ \hline 45 \end{array}$$

MR Langer

$$\begin{array}{r} 25 \\ + 15 \\ \hline 40 \end{array}$$

Answer: MR Langer

C 4

The school board wants to buy a piece of land. Mrs. Gomez and Mr. Langer are each selling a rectangular piece of land next to the school.

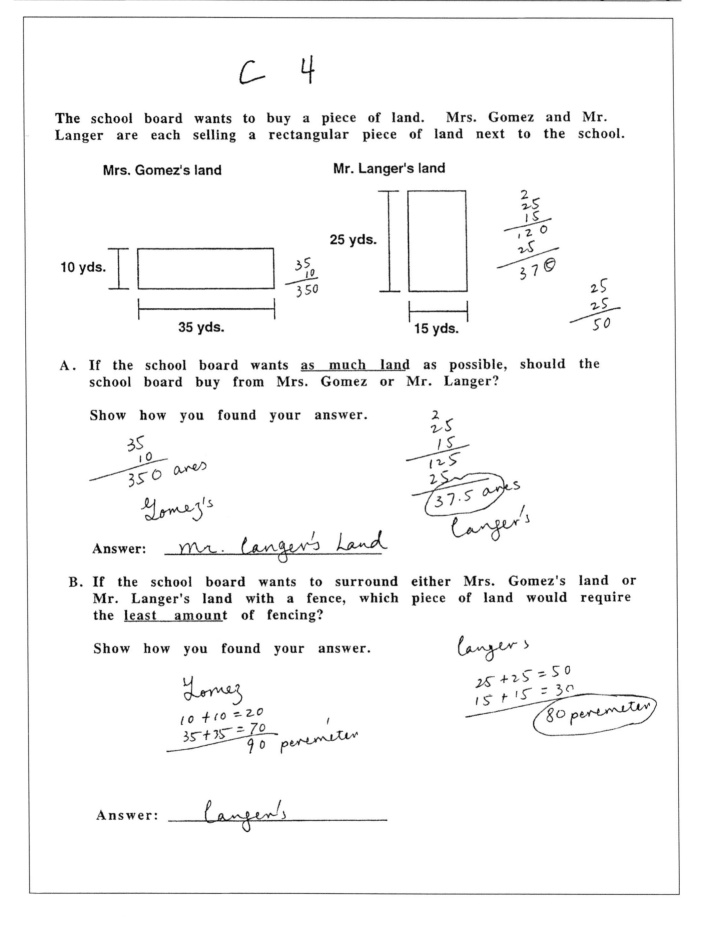

Mrs. Gomez's land

10 yds.

35 yds.

$\frac{35}{10}$
$\overline{350}$

Mr. Langer's land

25 yds.

15 yds.

$\begin{array}{r} 2 \\ 25 \\ 15 \\ \hline 120 \\ 25 \\ \hline 370 \end{array}$

$\begin{array}{r} 25 \\ 25 \\ \hline 50 \end{array}$

A. If the school board wants <u>as much land</u> as possible, should the school board buy from Mrs. Gomez or Mr. Langer?

Show how you found your answer.

$\begin{array}{r} 35 \\ 10 \\ \hline 350 \end{array}$ ares

Gomez's

$\begin{array}{r} 2 \\ 25 \\ 15 \\ \hline 125 \\ 25 \\ \hline 37.5 \end{array}$ ares

Langer's

Answer: <u>Mr. Langer's Land</u>

B. If the school board wants to surround either Mrs. Gomez's land or Mr. Langer's land with a fence, which piece of land would require the <u>least amount</u> of fencing?

Show how you found your answer.

Gomez
$10 + 10 = 20$
$35 + 35 = 70$
$\overline{90}$ perimeter

Langers
$25 + 25 = 50$
$15 + 15 = 30$
$\overline{80}$ perimeter

Answer: <u>Langer's</u>

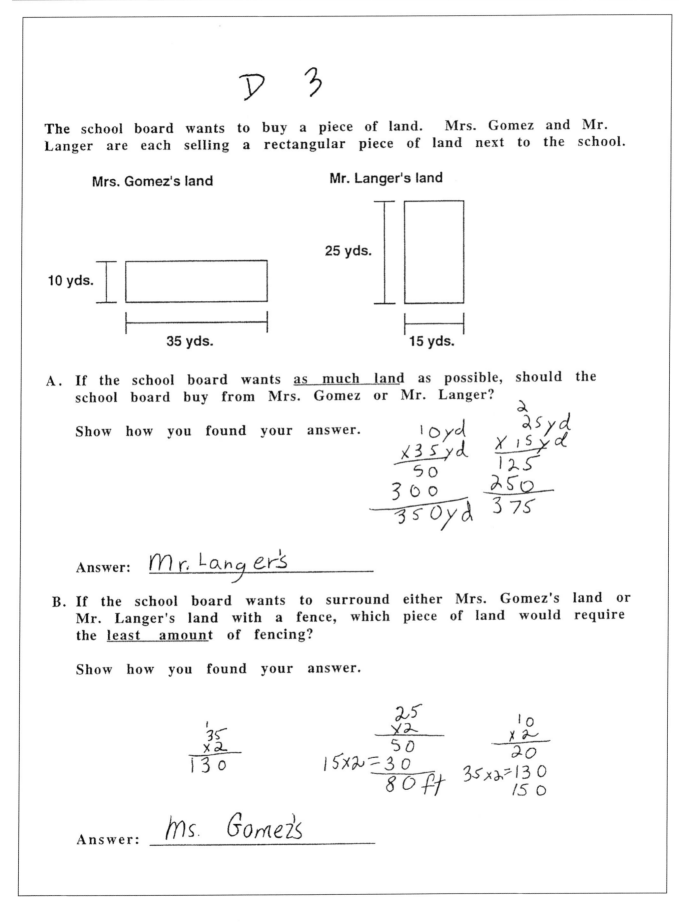

D 3

The school board wants to buy a piece of land. Mrs. Gomez and Mr. Langer are each selling a rectangular piece of land next to the school.

Mrs. Gomez's land   Mr. Langer's land

10 yds.

35 yds.

25 yds.

15 yds.

A. If the school board wants <u>as much land</u> as possible, should the school board buy from Mrs. Gomez or Mr. Langer?

   Show how you found your answer.

   10 yd
   x 35 yd
   ────
     50
   3 0 0
   ────
   3 5 0 yd

   25 yd
   x 1 5 yd
   ────
   1 2 5
   2 5 0
   ────
   3 7 5

   Answer: ___Mr. Langer's___

B. If the school board wants to surround either Mrs. Gomez's land or Mr. Langer's land with a fence, which piece of land would require the <u>least amount</u> of fencing?

   Show how you found your answer.

   35
   x 2
   ───
   1 3 0

   25
   x 2
   ───
    50
   15 x 2 = 3 0
   ────
   8 0 ft

   10
   x 2
   ───
    20
   35 x 2 = 1 3 0
   ────
   1 5 0

   Answer: ___Ms. Gomez's___

E 3

The school board wants to buy a piece of land. Mrs. Gomez and Mr. Langer are each selling a rectangular piece of land next to the school.

Mrs. Gomez's land

10 yds.

35 yds.

Mr. Langer's land

25 yds.

15 yds.

A. If the school board wants <u>as much land</u> as possible, should the school board buy from Mrs. Gomez or Mr. Langer?

Show how you found your answer.

$$35$$
$$10$$
$$00$$
$$35$$

$$\overset{2}{2}5$$
$$15$$
$$125$$
$$250$$
$$375$$

Answer: <u>35 / 350 Mrs Gomez</u>

B. If the school board wants to surround either Mrs. Gomez's land or Mr. Langer's land with a fence, which piece of land would require the <u>least amount</u> of fencing?

Show how you found your answer.

$$35 + 10 = 45$$
$$25 + 15 = 40$$

Answer: <u>Mr. Langer</u>

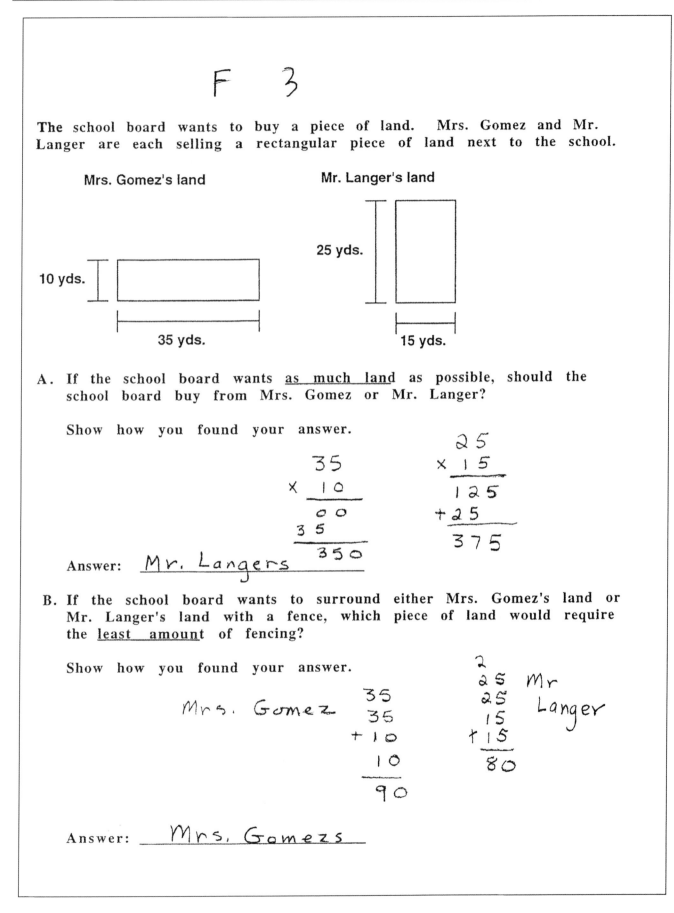

F 3

The school board wants to buy a piece of land. Mrs. Gomez and Mr. Langer are each selling a rectangular piece of land next to the school.

Mrs. Gomez's land

Mr. Langer's land

25 yds.

10 yds.

35 yds.

15 yds.

A. If the school board wants <u>as much land</u> as possible, should the school board buy from Mrs. Gomez or Mr. Langer?

Show how you found your answer.

$$\begin{array}{r} 35 \\ \times\ 10 \\ \hline 00 \\ 35\ \ \\ \hline 350 \end{array}$$

$$\begin{array}{r} 25 \\ \times\ 15 \\ \hline 125 \\ +25\ \ \\ \hline 375 \end{array}$$

Answer: <u>Mr. Langers</u>

B. If the school board wants to surround either Mrs. Gomez's land or Mr. Langer's land with a fence, which piece of land would require the <u>least amount</u> of fencing?

Show how you found your answer.

Mrs. Gomez
$$\begin{array}{r} 35 \\ 35 \\ +10 \\ 10 \\ \hline 90 \end{array}$$

$$\begin{array}{r} \overset{2}{25} \\ 25 \\ 15 \\ +15 \\ \hline 80 \end{array}$$ Mr Langer

Answer: <u>Mrs. Gomezs</u>

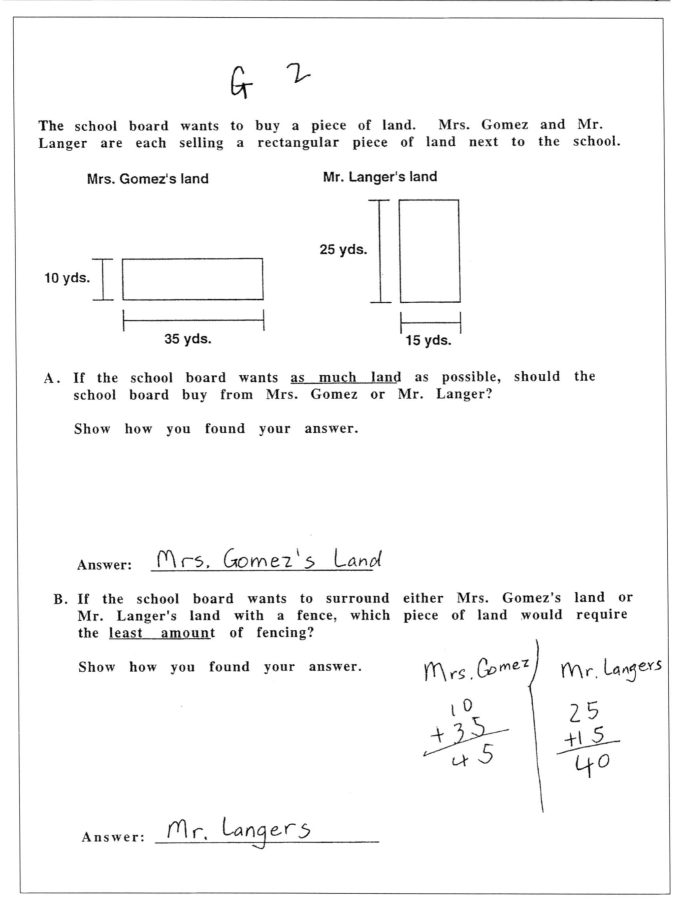

G 2

The school board wants to buy a piece of land. Mrs. Gomez and Mr. Langer are each selling a rectangular piece of land next to the school.

Mrs. Gomez's land   **Mr. Langer's land**

10 yds.   25 yds.

35 yds.   15 yds.

A. If the school board wants <u>as much land</u> as possible, should the school board buy from Mrs. Gomez or Mr. Langer?

Show how you found your answer.

Answer: <u>Mrs. Gomez's Land</u>

B. If the school board wants to surround either Mrs. Gomez's land or Mr. Langer's land with a fence, which piece of land would require the <u>least amount</u> of fencing?

Show how you found your answer.

Mrs. Gomez) Mr. Langers

$\begin{array}{r} 1\,0 \\ +3\,5 \\ \hline 4\,5 \end{array}$ $\begin{array}{r} 2\,5 \\ +1\,5 \\ \hline 4\,0 \end{array}$

Answer: <u>Mr. Langers</u>

H 2

The school board wants to buy a piece of land. Mrs. Gomez and Mr. Langer are each selling a rectangular piece of land next to the school.

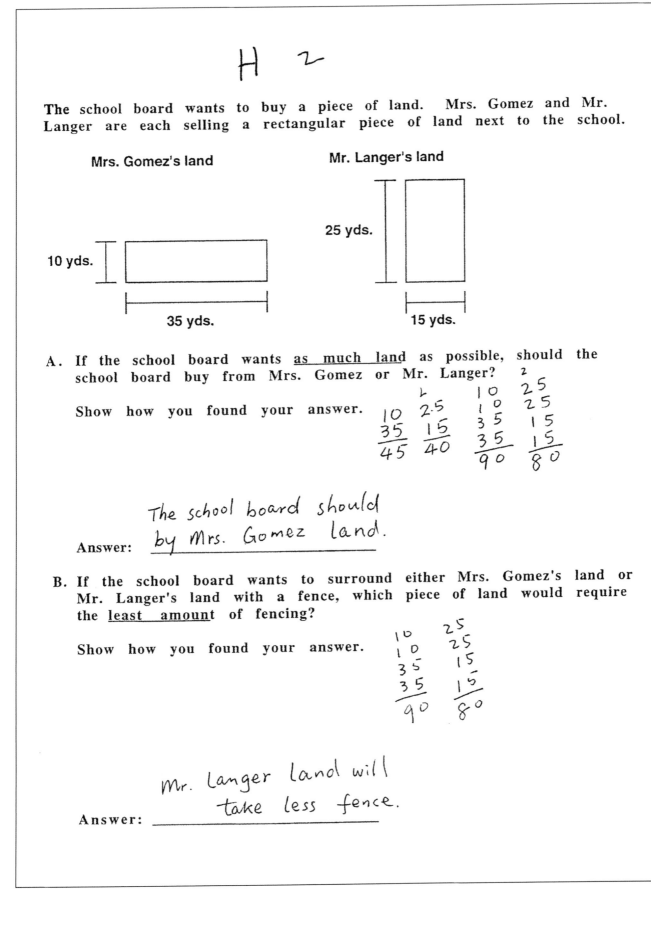

Mrs. Gomez's land   Mr. Langer's land

25 yds.

10 yds.

35 yds.   15 yds.

A. If the school board wants <u>as much land</u> as possible, should the school board buy from Mrs. Gomez or Mr. Langer?

Show how you found your answer.

```
    2
  L        1 0    2 5
10   2.5   1 0    2 5
35   15    3 5    1 5
45   40    3 5    1 5
          ─────  ─────
           9 0    8 0
```

Answer: The school board should by Mrs. Gomez land.

B. If the school board wants to surround either Mrs. Gomez's land or Mr. Langer's land with a fence, which piece of land would require the <u>least amount</u> of fencing?

Show how you found your answer.

```
1 0    2 5
1 0    2 5
3 5    1 5
3 5    1 5
─────  ─────
 9 0    8 0
```

Answer: Mr. Langer land will take less fence.

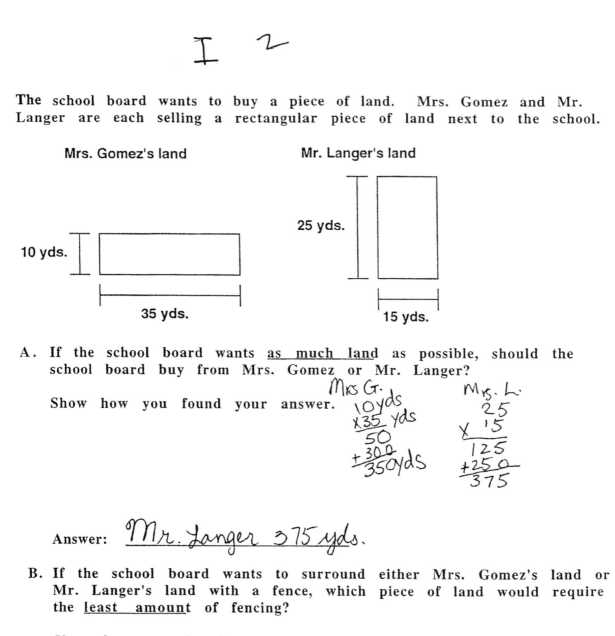

I 2

The school board wants to buy a piece of land. Mrs. Gomez and Mr. Langer are each selling a rectangular piece of land next to the school.

Mrs. Gomez's land   **Mr. Langer's land**

10 yds.   25 yds.

35 yds.   15 yds.

A. If the school board wants <u>as much land</u> as possible, should the school board buy from Mrs. Gomez or Mr. Langer?

Show how you found your answer.

Mrs G.
```
 10yds
x35 yds
  50
+300
350yds
```

Mrs. L.
```
  25
x 15
 125
+250
 375
```

Answer: _Mr. Langer 375 yds._

B. If the school board wants to surround either Mrs. Gomez's land or Mr. Langer's land with a fence, which piece of land would require the <u>least amount</u> of fencing?

Show how you found your answer.

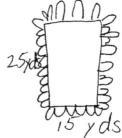

25yds
15 yds

Answer: _37 pieces of fencing_

J 1

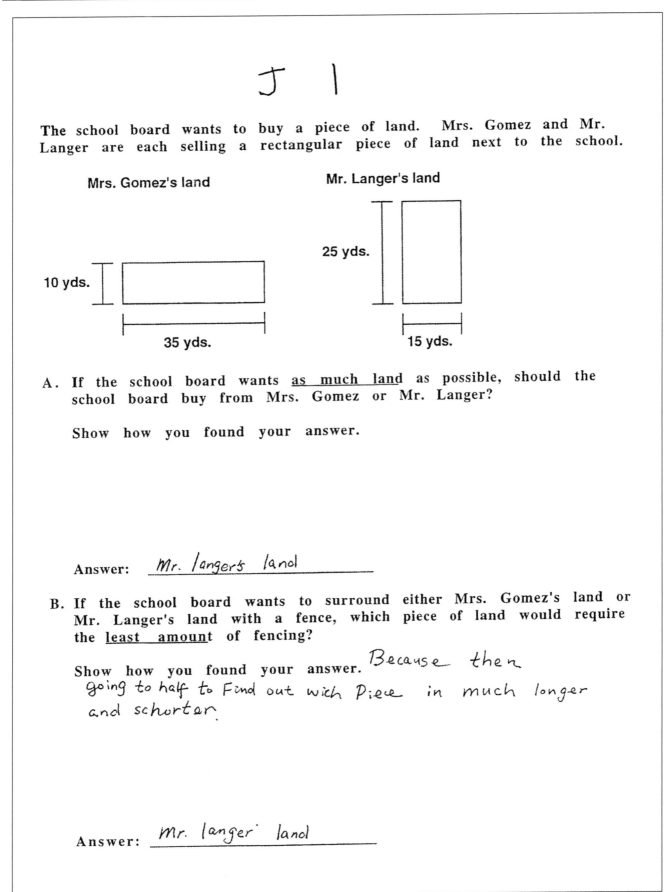

The school board wants to buy a piece of land. Mrs. Gomez and Mr. Langer are each selling a rectangular piece of land next to the school.

Mrs. Gomez's land

Mr. Langer's land

10 yds.

35 yds.

25 yds.

15 yds.

A. If the school board wants <u>as much land</u> as possible, should the school board buy from Mrs. Gomez or Mr. Langer?

Show how you found your answer.

Answer: _Mr. langer's land_

B. If the school board wants to surround either Mrs. Gomez's land or Mr. Langer's land with a fence, which piece of land would require the <u>least amount</u> of fencing?

Show how you found your answer. _Because then going to half to Find out wich Piece in much longer and schortor._

Answer: _mr. langer land_

K 1

The school board wants to buy a piece of land. Mrs. Gomez and Mr. Langer are each selling a rectangular piece of land next to the school.

Mrs. Gomez's land   Mr. Langer's land

25 yds.

10 yds.

35 yds.   15 yds.

A. If the school board wants <u>as much land</u> as possible, should the school board buy from Mrs. Gomez or Mr. Langer?

Show how you found your answer.

```
 10    35    For bot lands I        25      15
 10    35    Multiply each side     25      15
 ——    ——    2. and and them      ——     ——
 20    70    together              50      30
       90                                  
```
                                        80

Answer: ___Mrs. Gomez's land___

B. If the school board wants to surround either Mrs. Gomez's land or Mr. Langer's land with a fence, which piece of land would require the <u>least amount</u> of fencing?

Show how you found your answer.

Answer: ___Mr. Langer's land___

L I

The school board wants to buy a piece of land. Mrs. Gomez and Mr. Langer are each selling a rectangular piece of land next to the school.

Mrs. Gomez's land   Mr. Langer's land

25 yds.

10 yds.

35 yds.   15 yds.

A. If the school board wants <u>as much land</u> as possible, should the school board buy from Mrs. Gomez or Mr. Langer?

Show how you found your answer. because Mr. Langer Land is kind of squarish and Mr. Gomez Land is reqtangular

Answer: <u>Mr. Langers Land</u>

B. If the school board wants to surround either Mrs. Gomez's land or Mr. Langer's land with a fence, which piece of land would require the <u>least amount</u> of fencing?

Show how you found your answer. because his Land is thicker and you don't have to go ᵗᵒ wide strechs of land

Answer: <u>Mr Gomez's Land</u>

M   O

The school board wants to buy a piece of land. Mrs. Gomez and Mr. Langer are each selling a rectangular piece of land next to the school.

Mrs. Gomez's land   **Mr. Langer's land**

10 yds. — 35 yds.

25 yds. — 15 yds.

A. If the school board wants <u>as much land</u> as possible, should the school board buy from Mrs. Gomez or Mr. Langer?

Show how you found your answer.

Answer: ___Mrs. Gomez___

B. If the school board wants to surround either Mrs. Gomez's land or Mr. Langer's land with a fence, which piece of land would require the <u>least amount</u> of fencing?

Show how you found your answer.

Answer: ___Mr. Langer___

N   O

The school board wants to buy a piece of land. Mrs. Gomez and Mr. Langer are each selling a rectangular piece of land next to the school.

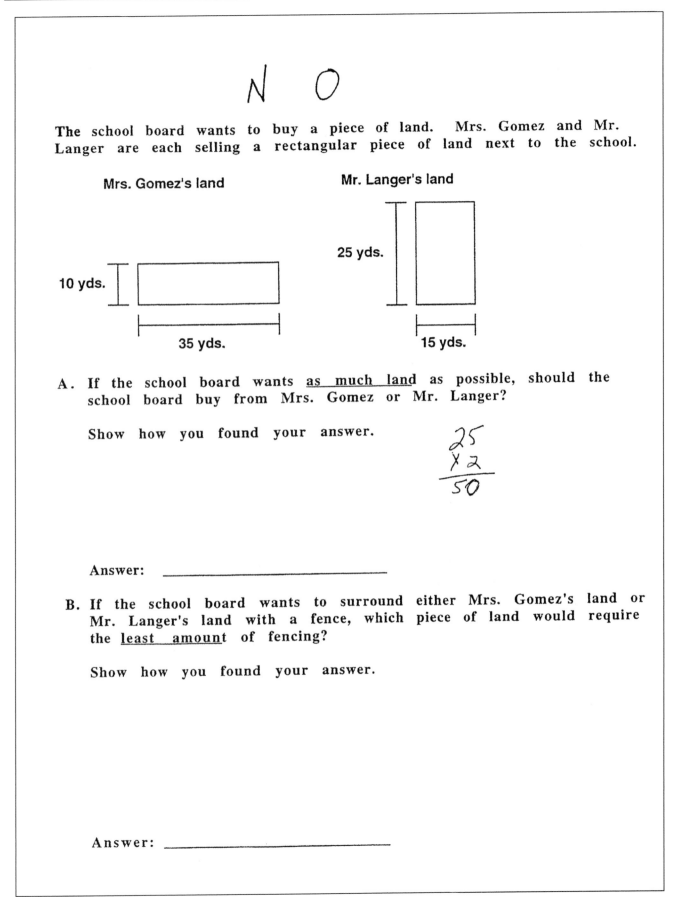

Mrs. Gomez's land

10 yds.

35 yds.

Mr. Langer's land

25 yds.

15 yds.

A. If the school board wants <u>as much land</u> as possible, should the school board buy from Mrs. Gomez or Mr. Langer?

Show how you found your answer.

$$\begin{array}{r} 25 \\ \times\ 2 \\ \hline 50 \end{array}$$

Answer: _____

B. If the school board wants to surround either Mrs. Gomez's land or Mr. Langer's land with a fence, which piece of land would require the <u>least amount</u> of fencing?

Show how you found your answer.

Answer: _____

P   o

The school board wants to buy a piece of land. Mrs. Gomez and Mr. Langer are each selling a rectangular piece of land next to the school.

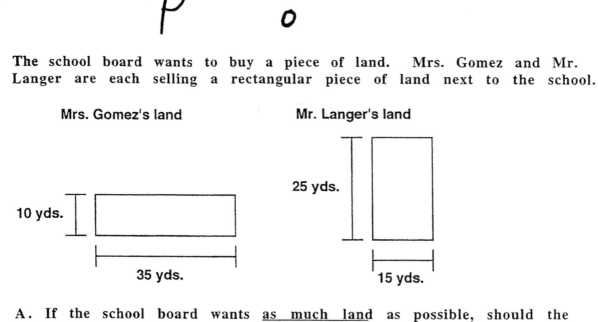

Mrs. Gomez's land

10 yds.

35 yds.

Mr. Langer's land

25 yds.

15 yds.

A. If the school board wants <u>as much land</u> as possible, should the school board buy from Mrs. Gomez or Mr. Langer?

Show how you found your answer.

$M r = 45$ $M_r = 40$

Answer: ___Mr Gomez___

B. If the school board wants to surround either Mrs. Gomez's land or Mr. Langer's land with a fence, which piece of land would require the <u>least amount</u> of fencing?

Show how you found your answer.

Same anser

Answer: ___Mr Langers___

DOUBLE THE CARPET

Double the Carpet Task

Mathematical Content

Measurement and geometry

Task Description

This task requires students to determine and compare the areas of two figures. The dimensions of two rooms are provided in a problem situation. The perimeter of the game room is double that of the living room. Students are asked to determine the correctness of the assertion that the "area to be carpeted in the game room is double the area to be carpeted in the living room." Students must describe how they made this decision. Arithmetic calculations, written explanations, and diagrams may be used to show that the area of the game room is not double the area of the living room but, rather, four times as large.

Parke, Carol S., Suzanne Lane, Edward A. Silver, and Maria E. Magone. *Using Assessment to Improve Middle-Grades Mathematics Teaching and Learning: Suggested Activities Using QUASAR Tasks, Scoring Criteria, and Students' Work*. Reston, Va.: National Council of Teachers of Mathematics, 2003.

Double the Carpet Task

Mr. Goldstein wants to buy carpet to cover the floors completely in his living room and game room. His living room is 10 feet by 15 feet. His game room is 20 <u>feet by</u> 30 feet.

Mr. Goldstein thinks that the area to be carpeted in his game room is double the area to be carpeted in his living room.

Is Mr. Goldstein correct?

> Explain your answer, and show your work. You may use diagrams as part of your explanation and work.

DOUBLE THE CARPET SCORING CRITERIA

Level 4

Explanation, work, or diagram completely and correctly compares the areas of two rectangular shapes when their dimensions are given. The fact that the area is not doubled is supported by work showing that the area of the game room is actually four times the area of the living room or that the living room fits into a quarter of the game room.

Level 3

Explanation, work, or diagram compares the areas of two rectangular shapes but contains a minor error or omission. For instance, a correct answer is given and the two areas are calculated correctly, but no explicit comparison between them is made.

Level 2

Explanation, work, or diagram indicates some comparison of the areas of the rectangular shapes. A correct or incorrect answer may be given.

Level 1

Explanation, work, or diagram shows a beginning understanding of the area concept; however, that understanding is limited. For instance, the area of only the living room is calculated.

Level 0

Explanation, work, or diagram reveals no understanding of the area concept. Work may calculate perimeters instead of areas.

Rationales for Scored Student Responses to Double the Carpet Task

Label	Score	Rationale
A	4	Correct answer of "no" is supported. Diagram and explanation completely and correctly show that the game room is actually four times as large as the living room.
B	4	Correct answer is supported. Calculations completely and correctly show that four times as much carpet is needed for the game room.
C	4	Correct answer is supported. A complete and correct diagram indicates that the living room fits into a quarter of the game room.
D	3	Correct answer is supported. Explanation is complete and compares the two areas; however, the student makes a calculation error in finding the area of the living room.
E	3	Correct answer is supported. The two calculations for the area are correct, but no explicit comparison between them is made.
F	3	Correct answer is supported. The diagram is correct, but the explanation contains an error. The size of the game room is not "3 times" that of the living room.
G	2	Correct answer is partially supported. Only the computation for the area of the game room is shown, and the explanation is incomplete.
H	2	Correct answer is partially supported. The diagram and explanation are incomplete. The comparison between the two areas is vague.
I	2	Incorrect answer is supported. The two area calculations are correct, but the "doubling" concept appears to be misinterpreted.
J	1	No answer is supported. Response correctly calculates the area of the living room and doubles it, but the dimensions for the game room are not considered.
K	1	No answer is supported. An attempt is made to calculate the area of the game room.
L	1	Incorrect answer is supported. Diagrams for both rooms are shown and labeled correctly, but the sizes of the rooms are not in proportion to each other.
M	0	Incorrect answer is supported. Work and explanation incorrectly compare the linear measures.
N	0	No answer is supported. Work shows the perimeters, not the areas.
P	0	Response shows inappropriate calculations.

4

Mr. Goldstein wants to buy carpet to cover the floors completely in is living room and game room. His living room is 10 feet by 15 feet. lis game room is 20 feet by 30 feet.

Mr. Goldstein thinks that the area to be carpeted in his game room is <u>ouble</u> the area to be carpeted in his living room.

Mr. Goldstein correct?

Explain your answer and show your work. You may use diagrams as part of your explanation and work.

living room

15 feet

feet

game room

15 feet + 15 feet = 30

10 feet   living room   living room
+
10 feet   living room   living room
= 20

Explain: Mr. goldstein is wrong The game room carpeted area is four times the size of the living room, it would be double The area if you only doubled one of The dimensions as you can see in my sketch.

4   B

Mr. Goldstein wants to buy carpet to cover the floors completely in his living room and game room. His living room is 10 feet by 15 feet. His game room is 20 feet by 30 feet.

Mr. Goldstein thinks that the area to be carpeted in his game room is <u>double</u> the area to be carpeted in his living room.

Is Mr. Goldstein correct?

> Explain your answer and show your work. You may use diagrams as part of your explanation and work.

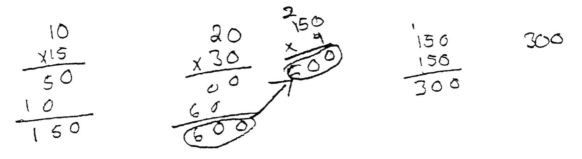

No it WAS 4 times As more cARpet For the gAme Room.

4   C

Mr. Goldstein wants to buy carpet to cover the floors completely in his living room and game room. His living room is 10 feet by 15 feet. His game room is 20 feet by 30 feet.

Mr. Goldstein thinks that the area to be carpeted in his game room is <u>double</u> the area to be carpeted in his living room.

Is Mr. Goldstein correct?

Explain your answer and show your work. You may use diagrams as part of your explanation and work.

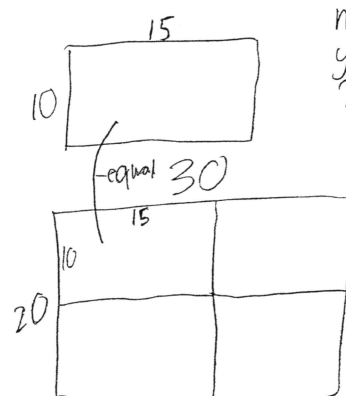

no he is not because you x 4 und get 20 x 30 carpet

3   D

Mr. Goldstein wants to buy carpet to cover the floors completely in his living room and game room. His living room is 10 feet by 15 feet. His game room is 20 feet by 30 feet.

Mr. Goldstein thinks that the area to be carpeted in his game room is <u>double</u> the area to be carpeted in his living room.

Is Mr. Goldstein correct?

Explain your answer and show your work. You may use diagrams as part of your explanation and work.

$$
\begin{array}{r} 10 \\ \underline{15} \\ 50 \\ \underline{20} \\ 250 \end{array}
\qquad
\begin{array}{r} 20 \\ \underline{30} \\ 00 \\ \underline{600} \\ 600 \end{array}
$$

$$
\begin{array}{r} 250 \\ \underline{2} \\ 500 \end{array}
$$

The total area of the living room is 250 because it is 10×15. The whole area of the game room is 600. 250×2 (which is double) equals 500 not 600

3   E

Mr. Goldstein wants to buy carpet to cover the floors completely in his living room and game room. His living room is 10 feet by 15 feet. His game room is 20 feet by 30 feet.

Mr. Goldstein thinks that the area to be carpeted in his game room is <u>double</u> the area to be carpeted in his living room.

Is Mr. Goldstein correct? No

Explain your answer and show your work. You may use diagrams as part of your explanation and work.

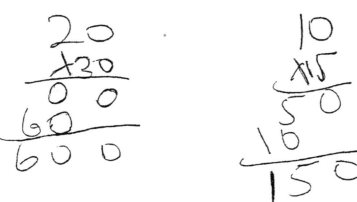

3   F

Mr. Goldstein wants to buy carpet to cover the floors completely in his living room and game room. His living room is 10 feet by 15 feet. His game room is 20 feet by 30 feet.

Mr. Goldstein thinks that the area to be carpeted in his game room is <u>double</u> the area to be carpeted in his living room.

Is Mr. Goldstein correct?

Explain your answer and show your work. You may use diagrams as part of your explanation and work.

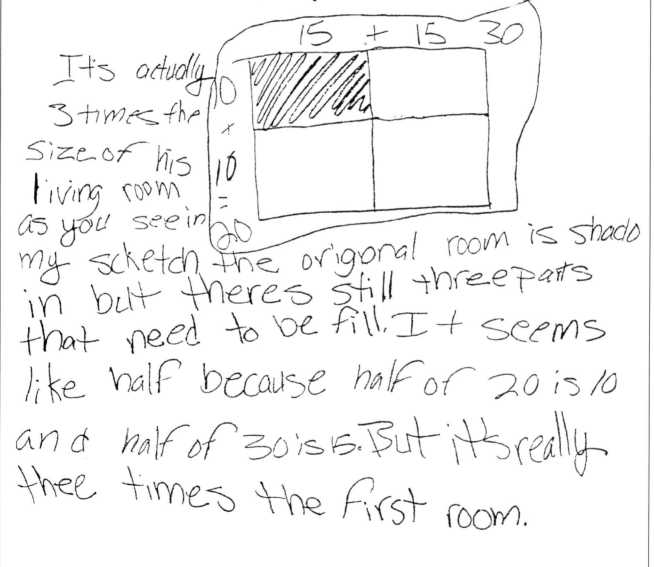

It's actually 3 times the size of his living room as you see in my scketch the origonal room is shado in but theres still three parts that need to be fill. It seems like half because half of 20 is 10 and half of 30 is 15. But it's really thee times the first room.

2   G

Mr. Goldstein wants to buy carpet to cover the floors completely in his living room and game room. His living room is 10 feet by 15 feet. His game room is 20 feet by 30 feet.

Mr. Goldstein thinks that the area to be carpeted in his game room is <u>double</u> the area to be carpeted in his living room.

Is Mr. Goldstein correct?   *NO*

Explain your answer and show your work. You may use diagrams as part of your explanation and work.

He is not correct because his game room is a lot bigger than his living room.

20 x 30 = 600

2   H

Mr. Goldstein wants to buy carpet to cover the floors completely in his living room and game room. His living room is 10 feet by 15 feet. His game room is 20 feet by 30 feet.

Mr. Goldstein thinks that the area to be carpeted in his game room is double the area to be carpeted in his living room.

Is Mr. Goldstein correct? ~~Yes~~ no

Explain your answer and show your work. You may use diagrams as part of your explanation and work. the living room is less than 1/2

15

10

30

20

2   I

Mr. Goldstein wants to buy carpet to cover the floors completely in his living room and game room. His living room is 10 feet by 15 feet. His game room is 20 feet by 30 feet.

Mr. Goldstein thinks that the area to be carpeted in his game room is <u>double</u> the area to be carpeted in his living room.

Is Mr. Goldstein correct?

Explain your answer and show your work. You may use diagrams as part of your explanation and work.

Yes because 20 x 30 = 600   10 x 15 = 150

and you a 150 plus 150 and you answer is 300. you add 300 plus 300 and your answer is 600 so there for it is double the carpet in the living room.

1   J

Mr. Goldstein wants to buy carpet to cover the floors completely in his living room and game room. His living room is 10 feet by 15 feet. His game room is 20 feet by 30 feet.

Mr. Goldstein thinks that the area to be carpeted in his game room is <u>double</u> the area to be carpeted in his living room.

Is Mr. Goldstein correct?

   Explain your answer and show your work. You may use diagrams as part of your explanation and work.

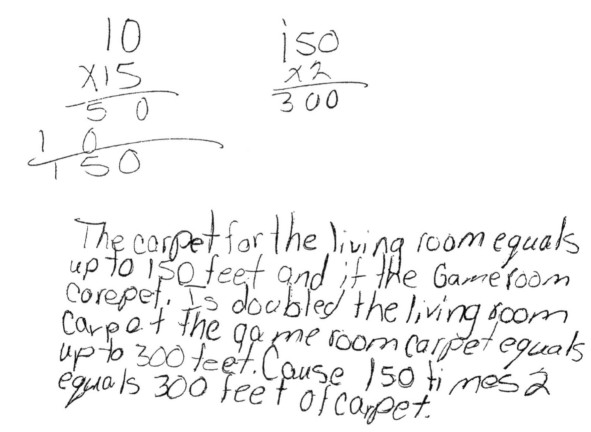

$$\begin{array}{r} 10 \\ \times 15 \\ \hline 50 \\ 10 \\ \hline 150 \end{array} \qquad \begin{array}{r} 150 \\ \times 2 \\ \hline 300 \end{array}$$

The carpet for the living room equals up to 150 feet and if the Gameroom corepet. Is doubled the living room Carpet the game room carpet equals up to 300 feet. Cause 150 times 2 equals 300 feet of carpet.

1   K

Mr. Goldstein wants to buy carpet to cover the floors completely in his living room and game room. His living room is 10 feet by 15 feet. His game room is 20 feet by 30 feet.

Mr. Goldstein thinks that the area to be carpeted in his game room is <u>double</u> the area to be carpeted in his living room.

Is Mr. Goldstein correct?

Explain your answer and show your work. You may use diagrams as part of your explanation and work.

$$\begin{array}{r} 20 \\ +30 \\ \hline 50 \end{array}$$

$$\begin{array}{r} 20 \\ \times 30 \\ \hline 00 \\ 06 \\ \hline 060 \end{array}$$

I L

Mr. Goldstein wants to buy carpet to cover the floors completely in his living room and game room. His living room is 10 feet by 15 feet. His game room is 20 feet by 30 feet.

Mr. Goldstein thinks that the area to be carpeted in his game room is <u>double</u> the area to be carpeted in his living room.

Is Mr. Goldstein correct?

Explain your answer and show your work. You may use diagrams as part of your explanation and work.

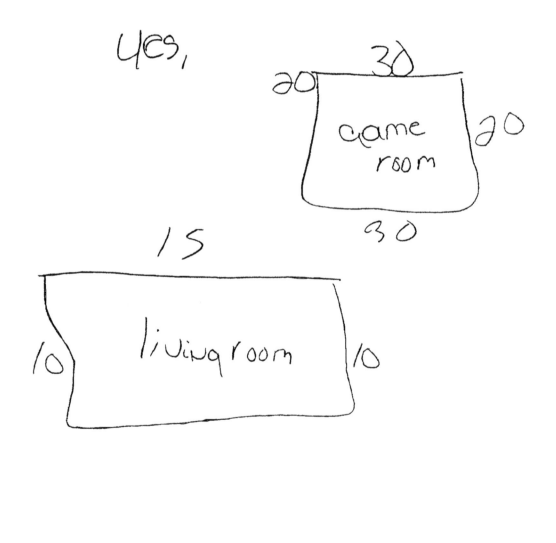

O   M

Mr. Goldstein wants to buy carpet to cover the floors completely in
his living room and game room. His living room is 10 feet by 15 feet.
His game room is 20 feet by 30 feet.

Mr. Goldstein thinks that the area to be carpeted in his game room is
<u>double</u> the area to be carpeted in his living room.

Is Mr. Goldstein correct? *yes*

Explain your answer and show your work. You may use
diagrams as part of your explanation and work.

See the living room is 10 feet by 15 feet &
his game room is 20 feet by 30 feet & 20 is double
10 & 30 is double 15.

$$\begin{array}{r} 20 \\ -10 \\ \hline 10 \end{array} \qquad \begin{array}{r} 30 \\ -15 \\ \hline 15 \end{array}$$

O

N

Mr. Goldstein wants to buy carpet to cover the floors completely in his living room and game room. His living room is 10 feet by 15 feet. His game room is 20 feet by 30 feet.

Mr. Goldstein thinks that the area to be carpeted in his game room is <u>double</u> the area to be carpeted in his living room.

Is Mr. Goldstein correct?

Explain your answer and show your work. You may use diagrams as part of your explanation and work.

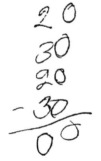

$$
\begin{array}{r}
20 \\
30 \\
20 \\
-\ 30 \\
\hline
00
\end{array}
$$

$$
\begin{array}{r}
10 \\
15 \\
10 \\
15 \\
\hline
50
\end{array}
$$

"O"   P

Mr. Goldstein wants to buy carpet to cover the floors completely in
his living room and game room.   His living room is 10 feet by 15 feet.
His game room is 20 feet by 30 feet.

Mr. Goldstein thinks that the area to be carpeted in his game room is
<u>double</u> the area to be carpeted in his living room.

Is Mr. Goldstein correct?   no

Explain your answer and show your work.   You may use
diagrams as part of your explanation and work.

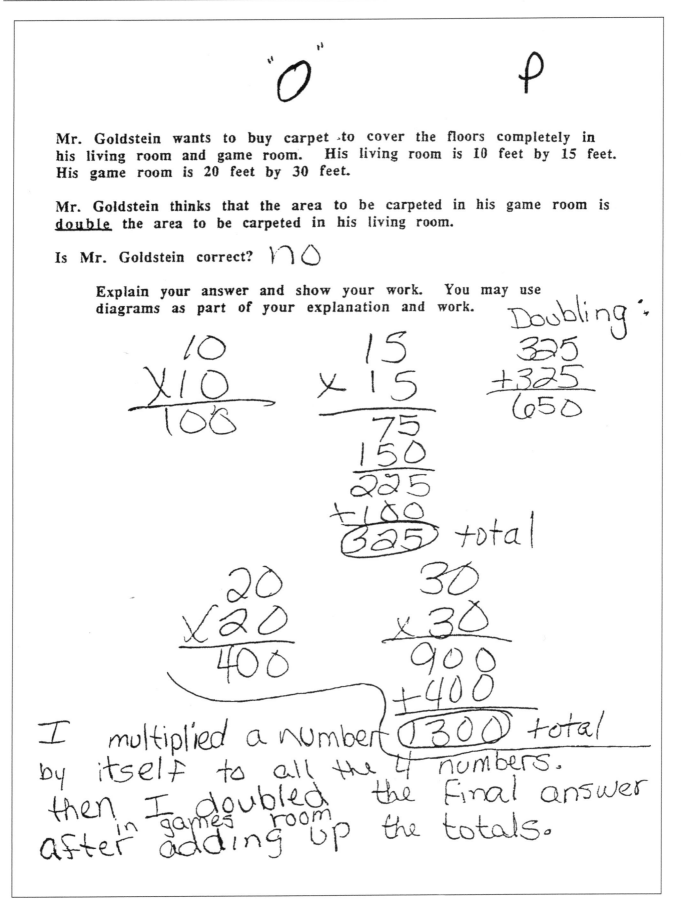

Doubling:
$$\begin{array}{r} 325 \\ +325 \\ \hline 650 \end{array}$$

$$\begin{array}{r} 10 \\ \times 10 \\ \hline 100 \end{array}$$

$$\begin{array}{r} 15 \\ \times 15 \\ \hline 75 \\ 150 \\ \hline 225 \\ +100 \\ \hline \boxed{325} \text{ total} \end{array}$$

$$\begin{array}{r} 20 \\ \times 20 \\ \hline 400 \end{array}$$

$$\begin{array}{r} 30 \\ \times 30 \\ \hline 900 \\ +400 \\ \hline \boxed{1300} \text{ total} \end{array}$$

I multiplied a number by itself to all the 4 numbers.
then I doubled the final answer
after adding up the totals.
in games room

APPENDIX

QUASAR Cognitive Assessment Instrument (QCAI) General Scoring Rubric

Score Level 4

Mathematical Knowledge: Shows understanding of the problem's mathematical concepts and principles; uses appropriate mathematical terminology and notations; and executes algorithms completely and correctly.

Strategic Knowledge: May use relevant outside information of a formal or informal nature; identifies all the important elements of the problem and shows understanding of the relationships between them; reflects an appropriate and systematic strategy for solving the problem; and gives clear evidence of a solution process, and solution process is complete and systematic.

Communication: Gives a complete response with a clear, unambiguous explanation and/or description; may include an appropriate and complete diagram; communicates effectively to the identified audience; presents strong supporting arguments which are logically sound and complete; may include examples and counterexamples.

Score Level 3

Mathematical Knowledge: Shows nearly complete understanding of the problem's mathematical concepts and principles; uses nearly correct mathematical terminology and notations; executes algorithms completely; and computations are generally correct but may contain minor errors.

Strategic Knowledge: May use relevant outside information of a formal or informal nature; identifies the most important elements of the problem and shows general understand-ing of the relationships between them; and gives clear evidence of a solution process, and solution process is complete or nearly complete, and systematic.

Communication: Gives a fairly complete response with reasonably clear explanations or descriptions; may include a nearly complete, appropriate diagram; generally communicates effectively to the identified audience; presents supporting arguments which are logically sound but may contain some minor gaps.

Score Level 2

Mathematical Knowledge: Shows understanding of some of the problem's mathematical concepts and principles; and may contain serious computational errors.

Strategic Knowledge: Identifies some important elements of the problems but shows only limited understanding of the relationships between them; and gives some evidence of a solution process, but solution process may be incomplete or somewhat unsystematic.

Communication: Makes significant progress towards completion of the problem, but the explanation or description may be somewhat ambiguous or unclear; may include a diagram which is flawed or unclear; communication may be somewhat vague or difficult to interpret; and arguments may be incomplete or may be based on a logically unsound premise.

Score Level 1

Mathematical Knowledge: Shows very limited understanding of the problem's mathematical concepts and principles; may misuse or fail to use mathematical terms; and may make major computational errors.

Strategic Knowledge: May attempt to use irrelevant outside information; fails to identify important elements or places too much emphasis on unimportant elements; may reflect an inappropriate strategy for solving the problem; gives incomplete evidence of a solution process; solution process may be missing, difficult to identify, or completely unsystematic.

Communication: Has some satisfactory elements but may fail to complete or may omit significant parts of the problem; explanation or description may be missing or difficult to follow; may include a diagram which incorrectly represents the problem situation, or diagram may be unclear and difficult to interpret.

Score Level 0

Mathematical Knowledge: Shows no understanding of the problem's mathematical concepts and principles.

Strategic Knowledge: May attempt to use irrelevant outside information; fails to indicate which elements of the problem are appropriate; copies part of the problem, but without attempting a solution.

Communication: Communicates ineffectively; words do not reflect the problem; may include drawings which completely misrepresent the problem situation.

Source: *QUASAR Cognitive Assessment Instrument (QCAI)* (Lane et al. 1995).
Used with permission.

REFERENCES

Curcio, Frances R. *A User's Guide to Japanese Lesson Study: Ideas for Improving Mathematics Teaching*. Reston, Va.: National Council of Teachers of Mathematics, 2002.

Kenney, Patricia A., Judi S. Zawojewski, and Edward A. Silver. "Marcy's Dot Pattern." *Mathematics Teaching in the Middle School* 3, no. 7 (1998): 474–77.

Lane, Suzanne. "The Conceptual Framework for the Development of a Mathematics Assessment Instrument for QUASAR." *Educational Measurement: Issues and Practice* 12 (1993): 16–23.

Lane, Suzanne, and Carol S. Parke. "The Consequences of a Performance Assessment in the Context of a Mathematics Instruction Reform Project." Paper presented at the annual meeting of the American Educational Research Association, San Francisco, California, 1995.

Lane, Suzanne, and Edward A. Silver. "Equity and Validity Considerations in the Design and Implementation of a Mathematics Performance Assessment: The Experience of the QUASAR Project." In *Equity and Excellence in Educational Testing and Assessment*, edited by Michael Nettles and Arie Nettles, pp. 185–219. Boston, Mass.: Kluwer Academic Publishers, 1995.

Lane, Suzanne, Edward A. Silver, Robert D. Ankenmann, Jinfa Cai, Connie Finseth, Mei Liu, Maria E. Magone, David Meel, Barbara Moskal, Carol S. Parke, Clement A. Stone, Ning Wang, and Yuehua Zhu. *QUASAR Cognitive Assessment Instrument (QCAI)*. Pittsburgh, Penn.: University of Pittsburgh, Learning Research and Development Center, 1995.

Lane, Suzanne, Clement A. Stone, Robert D. Ankenmann, and Mei Liu. "Reliability and Validity of a Mathematics Performance Assessment." *International Journal of Educational Research* 21 (1994): 247–66.

Magone, Maria E., Jinfa Cai, Edward A. Silver, and Ning Wang. "Validating the Cognitive Complexity and Content Quality of a Mathematics Performance Assessment." *International Journal of Educational Research* 21, no. 3 (1994): 317–40.

National Council of Teachers of Mathematics (NCTM). *Curriculum and Evaluation Standards for School Mathematics*. Reston, Va.: NCTM, 1989.

———. *Assessment Standards for School Mathematics*. Reston, Va.: NCTM, 1995.

———. *Principles and Standards for School Mathematics*. Reston, Va.: NCTM, 2000.

Parke, Carol S., and Suzanne Lane. "Learning from Performance Assessment in Math." *Educational Leadership* 54, no. 4 (1997): 26–29.

Silver, Edward A., Cengiz Alacaci, and Despina A. Stylianou. "Students' Performance on Extended Constructed-Response Tasks." In *Results from the Seventh Mathematics Assessment of the National Assessment of Educational Progress*, edited by Edward A. Silver and Patricia A. Kenney, pp. 301–41. Reston, Va.: National Council of Teacher of Mathematics, 2000.

Silver, Edward A., and Mary Kay Stein. "The QUASAR Project: The 'Revolution of the Possible' in Mathematics Instructional Reform in Urban Middle Schools." *Urban Education* 30, no. 4 (1996): 476–522.

Smith, Margaret Schwan. *Practice-Based Professional Development for Teachers of Mathematics*. Reston, Va.: National Council of Teachers of Mathematics, 2001.

Stein, Mary Kay, Margaret Schwan Smith, Majorie A. Henningsen, and Edward A. Silver. *Implementing Standards-Based Mathematics Instruction: A Casebook for Professional Development*. New York: Teachers College Press, and Reston, Va.: National Council of Teachers of Mathematics, 2000.

Stylianou, Despina A., Patricia A. Kenney, Edward A. Silver, and Cengiz Alacaci. "Gaining Insight into Students' Thinking through Assessment Tasks." *Mathematics Teaching in the Middle School* 6, no. 2 (2000): 136–44.